Important notice concerning the information in this book

This book is not intended to take the place of advice specific to your circumstances.

Information in this book is of a general nature only and may change over time. It does not purport to be a comprehensive statement of relevant regulations or guarantee any particular outcome. You should always exercise your own skill, care and judgement in using any information in this book. In particular, before using or relying on any information in this book you should carefully evaluate its currency, accuracy, completeness and relevance for your specific circumstances. You should seek legal and other professional advice on the requirements for your specific circumstances.

Always comply with manufacturers' instructions regarding the installation, use, maintenance and repair of any equipment of a type referred to in this book.

No liability will be taken for any error or omission or for any loss arising in connection with the use of any information in this book.

References (written or visual) to any organisation or to particular products or services are included for information or example only and should not be regarded as endorsements. The reference to any organisation, product or service in this book may not be used for advertising or product endorsement purposes.

Conversely, the fact that a particular organisation, product or service is not referenced should not be taken as any indication of any opinion of that organisation, product or service.

References to third-party websites and materials are provided for information and convenience only and not as endorsements. No responsibility is taken for the content and currency of any material included in third-party websites or materials.

Pumpkin Soup 660ml
Poppy Seed
Beigli
USE BY

Preface

The huge variety of manufactured foods available to Australian consumers today has largely been the result of the hard work and creativity of a group of relatively small manufacturers. Small businesses, defined as those employing less than 20 staff, make up approximately two-thirds of businesses in Australia's food and beverage manufacturing industry.

This book is aimed at those businesses currently manufacturing food on a small scale, or those considering entering this market. It will assist those already operating a small business to develop a better understanding of key food safety systems, while those who are in the 'start-up' phase will gain knowledge essential to providing their business with a solid foundation in food safety. The content will also be useful for students of food technology who wish to seek employment in the industry or are planning on establishing their own manufacturing operation.

It is often said by those associated with the food industry that food safety is non-negotiable, and this is the basis on which this book is written. Failure to control food safety hazards can have devastating consequences on the health of consumers and the viability of food businesses.

Information is presented in a straightforward, instructive manner, with details provided on the reasons why specific legislated food safety requirements are in place. Key messages are highlighted at the end of each chapter and provide a ready reference point for those seeking a summary of the main points.

The reader is led through a logical sequence of topics, starting with information on why food safety is so important to food business operators. Chapter 2 provides an outline of the three food safety hazards: microbial, chemical and physical. The control of microbial hazards is given special emphasis throughout the book because this is the greatest challenge to food manufacturers. Chapters 3 to 7 are devoted to practical guidance on how food safety hazards can be controlled. Areas covered include: premises, equipment, staff, product recipes, raw ingredients, preparation, processing, packaging, shelf-life and labelling. Chapter 8 covers the important topic of food recalls. Chapter 9 provides detailed information on each of the different types of pathogenic foodborne microorganisms of concern in Australia. To wrap up the book, the last sections – 'Sources of information', 'Useful contacts' and the 'Glossary' – provide a listing of references used in the book, contact details for those who can help small businesses with many aspects of food safety, and an explanation of terms in the book that the reader may not have encountered previously.

This book was written under the auspices of CSIRO Food and Nutritional Sciences (CFNS) and its predecessor, Food Science Australia (FSA). CFNS continues CSIRO and FSA's long history of providing food and nutrition research to support the health and wellbeing of the Australian community and the sustainability of the Australian food industry.

Sincere thanks go to Brigitte Cox and Keith Richardson who contributed significantly to the writing of this book. I would also like to thank other CFNS colleagues for their assistance:

- Heather Craven
- Patricia Desmarchelier
- Gary Dykes
- Vicki Eggelston
- Narelle Fegan
- Ailsa Hocking
- Atul Kacker
- Jennifer Keogh
- Cathy Moir
- Bob Steele

CFNS would also like to acknowledge and thank the representatives from the following bodies who contributed to the review process:

- Australian Capital Territory Department of Health
- Australian Food and Grocery Council
- Australian Government Department of Health and Ageing
- Coles Supermarkets
- New South Wales Food Authority
- Northern Territory Department of Health and Families
- Safe Food Production Queensland
- South Australian Department of Health
- TAFE New South Wales
- Tasmanian Department of Health and Human Services
- Victorian Department of Human Services
- Western Australian Department of Health, and
- Two representatives of the Australian small food business sector

CFNS received funding from the Australian Government towards this book.

Katherine Scurrah
Project Leader, CFNS

Contents

319 sick an... 105 in hospital –

local b...ness under scrutiny

She bit off more than she could chew - how Eliza lost her tooth!

...oisoning too

Cooling too slowly isn't cool!

10 ill after gourmet lunch

Vegetarians

...ys safe - bugs may ...uit and vegies

Peanuts in the ... leads to 7 year old's death

Raw eggs source of Salmonella

Hey! Who put that in my sausage?

Chapter 1

Why is food safety your concern?

Reducing the risk from food safety hazards is an essential part of any food production business – irrespective of its size. To help you achieve this, you need to know which food safety control measures are relevant to your products. If you are not sure what the terms 'risk', 'hazard' and 'control measure' mean, read Box 1.

Food safety is an issue that is taken very seriously by Australian government authorities. If you are manufacturing food products to sell, it is your responsibility to be aware of, and always follow, the food regulations relevant to the types of products you are producing. The overriding requirement is that food is 'safe' and 'suitable' or, in other words, it is fit for humans to eat. Information in this book will help you to ensure you achieve this outcome for your products.

What are the food safety hazards you should know about?

Below is a brief overview of the key food safety hazards: microbial, chemical and physical. More detailed information about each of these hazards is provided in Chapter 2.

Microbial hazards

Microbial foodborne illness, also commonly called 'food poisoning', is illness caused by eating food contaminated with specific types of microorganisms or toxins formed by these microorganisms. Microorganisms that are capable of causing illness are called 'pathogenic microorganisms' or simply 'pathogens'. Microorganisms that may be pathogenic are bacteria, viruses, parasites and moulds. There are many microorganisms that are not pathogenic but they can cause food to spoil (e.g. mould growth on bread). Although these spoilage microorganisms

Box 1 – Risks and hazards explained

People often get the meanings of the words 'risk' and 'hazard' mixed up or confused. Let's use an everyday example to explain these and another related term 'control measure'.

Definition of risks, hazards and associated terms

Term	Definition	Example
Hazard	Source of harm	Wet slippery floor (due to water leak)
Risk	Likelihood of being exposed to hazard and the likelihood of being harmed if exposed	Low risk – wet floor is in the corner of an unused room High risk – wet floor is in a poorly lit area that many people walk through
Control measures	The actions that can be taken to reduce the risk of exposure or fully eliminate the hazard	Reduce risk – wet floor warning sign is placed in front of area Eliminate hazard – leak is fixed and water is mopped up

are a concern for the food industry, they will not be discussed in any detail here because this book focuses on issues that may harm the health of consumers.

In Australia, the illness most commonly caused by eating food contaminated with pathogens or their toxins is foodborne gastroenteritis. The usual symptoms include nausea, diarrhoea and/or vomiting; these generally last for less than a week. Most people have either suffered from foodborne gastroenteritis at some stage of their lives or know someone who has.

There are some long-term illnesses that may develop after patients recover from symptoms of foodborne gastroenteritis. The effects of these can last for a much longer time than the initial gastroenteritis and may even cause lifetime disability. Irritable bowel syndrome – which causes abdominal pain, discomfort, bloating and alteration of bowel habits – is one example of this type of illness.

Some pathogens can cause illnesses other than foodborne gastroenteritis or gastroenteritis-associated illnesses. Examples of these are:

- flu-like symptoms

Box 2 – Vulnerable persons and food safety

The very young, the elderly and those whose immune system is seriously weakened have a higher chance of developing foodborne illness and may suffer more severe effects or complications. Weakening of the immune system can occur because of medication, such as chemotherapy treatments, or because of illness, such as AIDS. Pregnant women are also a vulnerable population as infection with pathogenic *Listeria monocytogenes* may cause miscarriages or stillbirths.

- damage to the nervous system
- damage to the liver
- damage to the kidneys
- meningitis – inflammation of the lining of the brain and spinal cord
- septicaemia – infection of the bloodstream
- encephalitis – inflammation of the brain.

Some people have a higher chance of contracting a foodborne illness than others, and they are also more likely to suffer more severely, develop more complications and, in some circumstance, die. These people are classed as 'vulnerable persons' (see Box 2 for further information).

Chemical hazards

In Australia, reports of illness caused by the presence of hazardous chemicals in food are much rarer than illnesses associated with pathogenic microorganisms. However, misuse of chemicals agriculturally or during food processing can cause illness. Harmful chemicals can also be present in food naturally or by environmental contamination. Additionally, substances found naturally in food can cause illness when eaten by people who are allergic or sensitive to them (Box 3).

Types and sources of chemicals in food that may cause illness if eaten include:

- proteins or other substances that may cause allergic reactions (e.g. peanuts)
- approved food additives, such as chemical preservatives, used incorrectly
- residues from cleaning and sanitising chemicals
- deliberate or accidental addition of chemicals not approved for use in food
- chemicals leaching from packaging into food

Box 3 – Food allergens

Food allergies are an exaggerated immune response by certain individuals to proteins or their derivatives that occur naturally in some foods. Foods most commonly reported to cause allergic reactions are peanuts, tree nuts, soy, milk, egg, cereals, fish, crustaceans and sesame.

In Australia, up to 8% of children and 2% of adults are allergic to one or more food (Food Industry Guide to Allergen Management and Labelling, see page 271 for details).

Allergic reactions to foods vary greatly: from mild gastrointestinal discomfort to skin rashes and potentially life-threatening breathing difficulties such as asthma and anaphylaxis. Even a tiny trace amount of an allergen can cause a severe reaction in sensitive people. The annual number of hospital admissions due to food-induced anaphylaxis has more than doubled over the past decade.

Many more people suffer from food intolerances rather than food allergies. Intolerances are reactions to natural or artificial substances found in some foods, such as lactose from cow's milk. Although not generally life-threatening (as food allergies can be), food intolerances can cause serious health problems. Coeliac disease is caused by intolerance to gluten, which is a protein present in wheat, oats, barley and rye. If people with the disease eat gluten, the lining of their small intestine may be damaged and they might be unable to absorb nutrients from food properly.

- environmental pollutants from industrial waste, such as mercury or dioxins in fish
- agrochemicals such as pesticide residues, herbicides and veterinary chemicals
- toxins of microbial origin such as mycotoxins in peanuts and histamines in fish
- naturally occurring plant toxins such as glycoalkaloids in potatoes.

Illnesses associated with chemicals in food can be caused by eating a high dose of a chemical contaminant over a short period (i.e. an acute reaction) or by eating a low level of a chemical contaminant over a long period of time (i.e. a chronic reaction).

Physical hazards

Physical contaminants in foods are objects that, under normal circumstances, should not be present in food products. Common contaminants include glass, bone, wood, metal, plastic, rubber, stones and insects.

Food safety concerns associated with physical contaminants include:

- choking (particularly young children)
- cuts to the mouth and tongue
- broken teeth
- damage to the gastrointestinal system.

Taking responsibility for food safety – it's the Law

In Australia, it is illegal to sell food that does not comply with relevant national standards or state and territory requirements. Even if you intend to export all of the products that you make, you must still meet national and state or territory requirements applicable to where the products are made, in addition to the requirements of the importing country.

The Australia New Zealand food standards system

The Australia New Zealand food standards system is a cooperative arrangement between the governments of Australia and New Zealand to develop and implement uniform food standards. This system operates under the Food Regulation Agreement 2002, and is implemented by food legislation in each state and territory, and by the *Food Standards Australia New Zealand Act 1991* (FSANZ Act) of the Commonwealth of Australia. Under the FSANZ Act, the Australian New Zealand Food Standards Code was developed.

The Australian New Zealand Food Standards Code is referred to as 'the Code' throughout the remainder of the book.

The Code is a collection of individual food standards, several of which contain specific requirements designed to reduce the risk of food businesses (Box 4) selling or supplying unsafe food. User guides are also available to help food businesses correctly implement the requirements of the Code. A list of the standards and user guides containing food safety regulations and related information is provided at the end of the book. All references to the Code in this book are based on standards current at the time of publication. Because standards can be updated or amended over time, it is the responsibility of food businesses to make sure they are aware of these changes (see page 269 for how you can keep up to date).

State and territory requirements

Responsibility for enforcing and ensuring compliance with the Code rests with regulatory authorities in each Australian state and territory. In most states and territories, local councils

Box 4 – What is a food business?

In the Code, a 'food business' is defined as:

A business, enterprise or activity, other than primary food production, that involves:
- the handling of food intended for sale, or
- the sale of food,

regardless of whether the business, enterprise or activity concerned is of a commercial, charitable or community nature or whether it involves the handling or sale of food on one occasion only.

'Handling of food' includes the making, manufacturing, producing, collecting, extracting, processing, storing, transporting, delivering, preparing, preserving, packing, cooking, thawing, serving or displaying of food.

Source: Standard 3.1.1 Interpretation and Application

play a major role in this process. It is the responsibility of food business owners to find out which authority is relevant to them and what requirements they have to meet. Details for contact points for each state and territory are provided at the end of the book.

It is a requirement of the Code (Standard 3.2.2 Food Safety Practices and General Requirements) that before commencing any food handling operations, all food businesses provide details about their business and proposed food handling activities to their relevant state or territory authority. This process is called 'notification'.

There are areas where requirements between state and territory authorities differ, some of these are minor and some are significant. Examples of areas where there may be differences include:
- requirements to be licensed or registered with the authority
- requirements to have a Food Safety Program or other documented food safety management system
- requirements for food handlers to wear hair coverings (see Box 5 for the definition of a food handler).

It is your responsibility to find out which specific requirements you must comply with, at both the national and state or territory level. All Australian food businesses must comply with the requirements of the Code.

Box 5 – Who is a food handler?

Food handlers are those who directly engage in the handling of food or handle surfaces likely to come into contact with food.

Handling of food includes the making, manufacturing, producing, collecting, extracting, processing, storing, transporting, delivering, preparing, preserving, packing, cooking, thawing, serving or displaying of food. It also includes staff who clean food contact surfaces and food processing equipment.

Food Safety Programs

A Food Safety Program (FSP) is a documented food safety management system. It requires a business to examine its operation and put controls into place to manage any significant hazards that are likely to occur.

In Australia, food businesses currently required to have a documented FSP or other food safety management system (approved by the relevant authority) include:

- food service businesses that serve food to vulnerable populations: Standard 3.3.1 Food Safety Programs for Food Service to Vulnerable Persons
- businesses that produce, harvest, process or manufacture bivalve molluscs, such as oysters: Standard 4.2.1 Primary Production and Processing Standard for Seafood
- businesses that process or manufacture ready-to-eat (see Box 6) poultry meat, such as smoked chicken and diced cooked chicken meat (for use in salads, sandwiches, and so on): Standard 4.2.2 Primary Production and Processing Standard for Poultry Meat
- businesses that process or manufacture ready-to-eat meat products, such as pâté, ham and fermented meats: Standard 4.2.3 Primary Production and Processing Standard for Meat
- businesses that produce, transport, process or manufacture dairy products: Standard 4.2.4 Primary Production and Processing Standard for Dairy Products.

Additionally, at the time of publication of this book, FSANZ was working on a standard to require businesses that perform certain catering activities for the general public to develop and implement an FSP.

It is also mandatory in some states and territories for some or all food businesses to have an effective FSP in place, even if they do not perform any of the above listed activities.

Box 6 – What is a ready-to-eat food?

The term 'ready-to-eat' is used for food that does not require cooking or re-heating by the consumer before eating. It may have already been cooked by the manufacturer or it may normally be eaten raw. Examples include dips, salads, pâté, sandwiches and slow cured meats (e.g. prosciutto).

Greater emphasis is placed on the need for these products to be prepared under strict hygienic conditions because there is no cooking step in the home of the consumer to kill any contaminating pathogens. In contrast, products such as raw meat generally will have a cooking step, allowing the levels of pathogens potentially present to be reduced.

It is a food business owner's responsibility to find out if they are required to have an FSP and if so, with which specific standard of the Code they must comply.

Food Safety Programs must:

- identify all potential food safety hazards
- identify where and how these hazards can be controlled
- include systems for monitoring adequacy of these controls
- include systems for implementing corrective actions if the controls fail; corrective actions are required to stop the cause of the failure happening again
- include systems for regular review of the FSP to ensure it is adequate
- include appropriate record keeping systems.

The amount and type of information needed in an FSP depends on the type and complexity of the food business and food safety hazards. For example, a business producing a low-risk product such as jam would need a simpler FSP than a business that prepares a variety of potentially higher risk fresh dips.

Development of an FSP has to be appropriate to the specific operating environment and products manufactured or prepared by an individual food business. Businesses can't simply copy an FSP from another business or download one from the internet. There are numerous tools available, such as templates and computer software to assist with this process (see page 277). The guidance provided in this book may assist you in the development of an effective FSP.

Ensuring compliance – Environmental Health Officers

State and territory food authorities and/or local councils are responsible for ensuring that food businesses comply with legislated requirements. Environmental Health Officers (EHO), sometimes called Authorised Officers, are the representatives of these bodies with which food businesses will have contact.

One of the roles of an EHO is to visit food businesses, examine their documentation and records, and observe their food handling practices. The frequency of EHO visits is variable; for example, if you prepare higher risk products such as ready-to-eat chilled foods you can expect to be visited more frequently than if you make lower risk foods such as jam. If the authorities receive a complaint about a food business – say from a concerned member of the public – an inspection by an EHO may occur.

If an EHO determines that a business is not complying with the relevant regulations, the owners may be subject to a number of steps in an enforcement process. Steps in this process may include:

- serving of an improvement notice – stating changes that must be made within a specified timeframe, such as repairing or replacing faulty equipment
- issuing of a prohibition order – specifying activities that must cease until a clearance certificate is issued and/or ordering the destruction or disposal of food items
- issuing of a penalty notice – small 'on-the-spot' style fines that are set for specific breaches
- prosecution – larger fines may be imposed by a court for serious breaches or for a failure to respond appropriately to other steps in the process such as not paying a penalty notice fine within the specified timeframe.

In some states or territories, food businesses that violate regulations are named publicly.

The benefits of taking food safety issues seriously

Instead of asking, 'What will it cost me if someone becomes ill as a result of eating my product?' or 'What would it mean to me if I ended up going out of business?' you should ask, 'How will implementing food safety management systems benefit my business?'

Implementing a food safety management system is an investment in both the future of your business and the reputation of Australia's food industry. A preventative approach (i.e. stopping a problem occurring in the first place) rather than reactive (i.e. 'mopping up the mess') is the most responsible and cost-effective way to handle food safety hazards. Should your business be unfortunate enough to be implicated in a foodborne illness outbreak (Box 7), your documented food safety management system may possibly assist your defence.

There are very few foodborne illness outbreaks each year from Australian commercially manufactured or processed foods. Of the 149 outbreaks recorded in 2007, only three were linked to foods in this category (see diagram below). This demonstrates, in part, that Australian food businesses producing manufactured foods are investing an appropriate amount of resources into establishing and maintaining effective food safety management systems. However, if these systems are not in place, or are allowed to become ineffective, outbreaks caused by processed or manufactured food have the potential to affect many people. This is because larger volumes of products are generally made by these businesses compared with other food preparation settings (e.g. food prepared in a private home) and they are distributed more widely.

As the majority of foodborne illness in Australia is caused by pathogenic microorganisms, these will be our primary focus in this chapter. Most of the facts and figures provided in the remainder of this chapter are from the Australian Government's 2006 report, The Annual Cost of Foodborne Illness in Australia. Although the data in this report was gathered in 2001 and 2002, it provides the most comprehensive information for Australia available at the time of publication. For more information on this report see page 273.

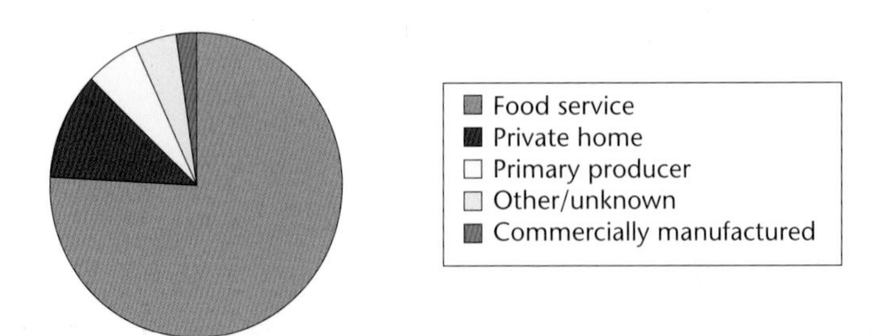

Food preparation settings implicated in foodborne illness outbreaks (2007). (Source: Monitoring the incidence and causes of diseases potentially transmitted by food in Australia: Annual Report of the OzFoodNet Network, 2007)

Box 7 – What is a foodborne illness outbreak?

When two or more individuals experience a similar illness after eating the same food type, or food from the same place, an outbreak has occurred. However, a large number of outbreaks go 'unnoticed', because those affected either do not visit a doctor or they visit a doctor but the true cause of their illness goes undiagnosed.

To determine the cause of a foodborne illness doctors need to take samples from the patient, which are sent to a laboratory for testing. Each Australian state and territory requires that illnesses caused by certain pathogens are notified to authorities. It is through this notification system that outbreaks are detected.

All outbreaks reported to authorities are investigated. This may involve:

- locating the source of the implicated food
- determining which particular food or foods were responsible
- determining what caused the food to become a food safety issue
- ensuring steps are taken to prevent future occurrences
- initiating a food recall, if necessary (Box 8).

The Australian Government collects data on foodborne illness outbreaks and analyses this for trends; for example, an increase in outbreaks associated with one particular food type. Guidance for the development of public education campaigns, changes to food regulations or other food safety related activities is then developed.

Health and wellbeing of your customers

There are over 5 million estimated cases of foodborne illness in Australia every year. This high burden of disease and the potential for those affected to suffer severe long-term illnesses – or even to die – is a reminder for food businesses to be constantly vigilant. This is a very significant problem, which must be treated seriously by anyone who handles food.

Estimation of foodborne illness in Australia for one year (2001–2002)

Illness	Number of cases	Number of deaths
Gastroenteritis	5.4 million	80
Other infections (non gastroenteritis)	6000	27
Long-term illness (post gastroenteritis)	41 000	18

Source: The Annual Cost of Foodborne Illness in Australia

Box 8 – Food recalls explained

A food recall is defined as 'an action taken to remove from sale, distribution and consumption foods that may pose an unacceptable safety risk to consumers'.

Recalls are initiated either by the food business (voluntary) or by government authorities (mandated).

Businesses or authorities initiate recalls for a variety of reasons including:

- the product may not meet the requirements of the Code
- complaints are received from consumers or retailers
- doctors or government agencies report cases of foodborne illness linked to the product.

Recalls may be performed at two levels:

- the trade level – which involves return of stock from distribution centres and wholesalers
- the consumer level – which also involves return of stock from retailers and consumers.

Recalls are publicised in a variety of ways such as notices in newspapers or signs posted in supermarkets to notify consumers.

More information about food recalls is provided in Chapter 8.

Viability and growth of your business

By taking action to reduce food safety risks, you can protect the health of your business by: building or maintaining a favourable reputation with your customers; avoiding costs associated with product recalls, and loss of sales or contracts; and avoiding additional financial costs such as legal liability.

A business lives on its reputation, and goodwill forms part of the value of the business. Most business owners are aware they need to develop a good reputation for quality and service, but a reputation for safety is even more important.

Larger companies, such as caterers or supermarkets purchasing from small businesses, need to be sure that they are purchasing food that is as safe as reasonably possible. These companies do not want to put their reputation at risk because of sub-standard food safety practices used by their suppliers. Consequently, most large businesses are now requiring their suppliers to implement a food safety management system, which they then audit (Box 9). Having a food safety management system, such as an FSP, can expand your business opportunities and help your business grow.

Food rarely becomes an unacceptable safety risk due to a random accident that is 'nobody's fault': the cause can usually be linked back to the failure of or, worse, complete lack of, food safety management systems at the implicated food business. Food recall statistics gathered by FSANZ show that failure to eliminate food safety hazards or to adequately reduce risks associated with hazards are common problems.

During 2008, there were 45 consumer-level recalls of food products in Australia; of these:

- 33% were due to physical contaminants
- 29% were due to microbial contamination – primarily the presence of foodborne pathogens
- 27% were due to incorrect labelling – primarily failure to warn about the presence of food allergens
- 11% were caused by other factors including chemical contaminants and processing failures.

Product recalls not only affect your reputation, they are also expensive exercises. Examples of expenses associated with a recall are:

- newspaper advertising; the scale depends on how many products are affected and how widely they are distributed (i.e. local area, nationwide or international)

- the cost of stock; includes refunds on products already sold and stock that has yet to be distributed or sold
- the recovery of stock; the cost depends on how widely the product is distributed, and whether the destruction of products will be performed at the point of sale or if all stock is to be returned to the manufacturer for disposal
- the destruction of stock; the cost depends on the amount of stock and the method required (e.g. normal trade waste or incineration)
- product testing costs; the cost depends on the type of testing required and the number of samples requiring testing
- other associated costs such as overtime payments, loss of profit due to indirect costs from disruption and loss of sales, and penalties from supermarkets for removal of stock from shelves.

The other significant financial costs you may face are fines imposed directly by government authorities or through the court system, or compensation payments. The following table shows cases of Australian businesses fined for incorrect practices or required to pay damages to victims of foodborne illness.

Fines paid by Australian food businesses (examples from 2006–2009)

Business type	Food safety issues (examples only, not complete listings)	Fines and/or legal costs
Retail outlet (chicken products)	Inadequate temperature control, inadequate pest control, unclean premises, inadequate monitoring of a Food Safety Program	$132 000
Sushi processor	Inadequate pest control, unhygienic practices (e.g. storing raw chicken in close contact with ready-to-eat foods)	$61 250
Café	Inadequate pest control, inadequately protecting food from contamination, unclean premises	$43 400
Bakery	Food stored and displayed at temperatures that allowed pathogenic bacteria to grow to dangerous levels	$42 000
Restaurant	Unhygienic practices (e.g. dirty appliances and food left uncovered), unclean premises, premises in disrepair	$25 000
Butcher	Illegal addition of food preservative	$11 970

Source: NSW Food Authority media releases, Victorian Government Food Safety News newsletter and WA Department of Health Notice of Convictions website

Continued popularity of products in your industry sector

Failing to take adequate responsibility for food safety in your business can have far-reaching effects on your industry sector. There are many businesses both in Australia and overseas that have not been able to survive after their products were associated with a foodborne illness outbreak. Even businesses not directly involved in the outbreak but producing the same type of food may suffer if there is a large amount of media coverage of the outbreak.

During 1994–1995, there was a foodborne illness outbreak linked to consumption of smallgoods manufactured by a South Australian (SA) business. Over 150 people became ill and one child died. The manufacturer went out of business within 2 months of the start of the outbreak. The manufacturer had been in business for 20 years, employed 120 people and had a yearly turnover of $13 million.

Sales of certain types of smallgoods products, both in SA and throughout Australia, were affected for several years after the outbreak (approximately 50% decrease in SA and 20%

Box 9 – What is an audit?

An audit is a structured assessment of a system to determine if it is capable of achieving its aims (i.e. does it make sense from a scientific or technical perspective?) and whether the procedures specified in the system are being followed. For example, if a supplier's food safety management system specifies that there must be a certain amount of a preservative in a product to ensure its safety, an auditor will look at several lots of batch records for these products to see if this requirement is consistently being met.

It is advisable that you conduct your own internal audits on your food safety management system because this helps to prepare for audits conducted by other parties, such as an EHO or companies you supply.

Auditors may do a 'desktop' audit, and just look at your documents and records or they may do a tour of your premises as well. You will receive a report that identifies areas that require improvement or you have overlooked. Auditors are not necessarily experts in your product, so don't be frightened to make your case, with documented evidence of course, if you believe the auditor is mistaken.

nationwide). In addition, it has been estimated that in the longer term this outbreak resulted in 400–500 smallgoods manufacturers going out of business.

Minimising the overall cost to society

It costs Australian taxpayers millions of dollars every year to fund government agencies responsible for maintaining Australia's safe food supply. This investment includes the cost of surveillance, investigating outbreaks and maintaining food safety management systems.

In addition, our public health-care systems, such as Medicare, bear large annual costs associated with foodborne illness. 'Productive days lost' is used to describe the number of days individuals suffering from a foodborne illness are unable to attend paid work or household duties. The

Estimate of health-care costs and lost work days associated with foodborne illness in Australia in one year (2001–2002)

Illness	Cost ($)	Productive days lost
Gastroenteritis	200 million	6 million
Other infections (non gastroenteritis)	3 million	18 000
Long-term illness (post gastroenteritis)	19 million	121 000

Source: The Annual Cost of Foodborne Illness in Australia

financial impact of these can be widespread. Costs include lost wages, child-care charges, cost to employers for hiring replacement staff and sickness benefits paid by government agencies.

Protecting the reputation of Australia's food industry

The Commonwealth Government reported that for the 2007–2008 financial year, the value of Australia's food exports was $23.4 billion and the processed food and beverage industry employed nearly 206 000 people (Australian Food Statistics 2008; see page 271).

The continued success of Australia's food export industry depends on maintenance of its reputation as a safe food producer. Just as the Australian Quarantine and Inspection Service (AQIS) has tight control over the safety and quality of food permitted into Australia, official bodies overseas set strict criteria on the food allowed into their jurisdictions. Any businesses planning to export their products must seek the assistance of AQIS to determine the regulatory requirements of the destination countries (see page 276 for AQIS contact details).

The occurrence of widely publicised food safety scares, such as those listed below, justifies the ongoing need for vigilance by Australian and international bodies:

- 2008 – A United States foodborne illness outbreak was linked to peanut butter (over 600 people became ill and over 2700 products recalled).
- 2008 – Chinese infant formula products were contaminated with the industrial chemical melamine (see Box 10).

Box 10 – Melamine contamination of Chinese infant formula

In China in 2008, a major foodborne illness outbreak was caused by intentional misuse of an industrial chemical, melamine.

Melamine – a chemical used in the production of plastic materials and adhesives – was added to milk used to make infant formula and other dairy products. This was done deliberately to make it appear as if the milk had much higher levels of protein than were actually present.

An estimated 300 000 infants and young children became ill because of this illegal adulteration. More than 50 000 of those affected were hospitalised for kidney stones or other urinary tract issues.

- 2006 – United Kingdom-manufactured chocolate bars were contaminated with pathogenic bacteria (1 million potentially affected products were recalled).
- 2006 – A United States foodborne illness outbreak was linked to spinach (this is estimated to have cost the Californian spinach industry over US$50 million).
- 1996 – An Australia-wide recall of peanut butter contaminated with pathogenic bacteria was conducted (this cost the manufacturer over $55 million).

Any internationally publicised food safety issue arising in Australia has the potential to make overseas markets wary, and may tarnish our food industry's reputation and consequently damage our economy.

KEY MESSAGES FROM CHAPTER 1

- Identifying and controlling food safety hazards is the responsibility of all food business operators.
- There are three categories of food safety hazards:
  - microbial (e.g. pathogenic bacteria)
  - chemical (e.g. food allergens)
  - physical (e.g. broken glass).
- There are over 5 million cases of foodborne illness in Australia yearly; gastroenteritis is the most common illness, costing our health-care system an estimated $200 million annually.
- The very young, the elderly, those with weakened immune systems and pregnant women have a higher chance of developing foodborne illness and suffering severe effects or complications.
- Severe allergic reactions can be caused by naturally occurring substances in food or by added substances.
- It is illegal to sell food in Australia that does not comply with relevant national standards, or state and territory requirements.
- The Australian New Zealand Food Standards Code is referred to as 'the Code' throughout the book.
- State and territory food authorities and/or local councils are responsible for ensuring food businesses comply with legal requirements; Environmental Health Officers audit businesses to monitor this and investigate consumer complaints.
- A preventative, rather than reactive, approach is the most responsible and cost-effective way to control food safety hazards.
- Product recalls not only affect a businesses reputation, they can be very costly.
- Many Australian businesses have not been able to survive after their products were associated with a foodborne illness outbreak.

CAUTION
KEEP OUT OF REACH OF CHILDREN
READ SAFETY DIRECTIONS
DOUBLE STRENGTH
KILLS RATS & MICE
SPRAY GREASE
LITHIUM BASE
APPLICATION
POISON

REMOVE

Chapter 2

Food safety hazards – under the spotlight

Chapter 1 contained a brief introduction to the three primary food safety hazards. More detail is provided in this chapter – the more you 'know your enemies' the better equipped you will be to stop them being a problem in your products.

As the most common cause of foodborne illness in humans is pathogenic microorganisms – particularly bacteria and viruses – much more detailed information will be provided on these compared with the other hazards.

Microbial hazards – pathogenic microorganisms and their toxins

A 'microorganism' is an organism that is usually too small to be seen without using a microscope (Box 11). Microorganisms that can be found in food include bacteria, viruses, yeasts, moulds, parasites and algae. These can be present in food simply because they are part of the natural environment. They can be intentionally introduced to food as part of a manufacturing process, such as during yoghurt production, or they can contaminate food during production, preparation or handling.

Viruses and pathogenic bacteria are the microorganisms most likely to cause isolated cases or outbreaks of foodborne illness in Australia. Some types of pathogenic bacteria produce toxins, and it is the toxins, and not the bacteria themselves, that cause illness. Parasites also cause foodborne illness cases in Australia, although these are rarer. Specific types of mould are capable of producing harmful chemical compounds, called mycotoxins, in food. Some algae can also

Box 11 – Just how small are microorganisms?

Below is a photo of a cluster of bacterial cells magnified 10 000 times. The scale on the photo shows 1 μm (one millionth of a metre). To put this into perspective, a strand of human hair is about 100 μm wide.

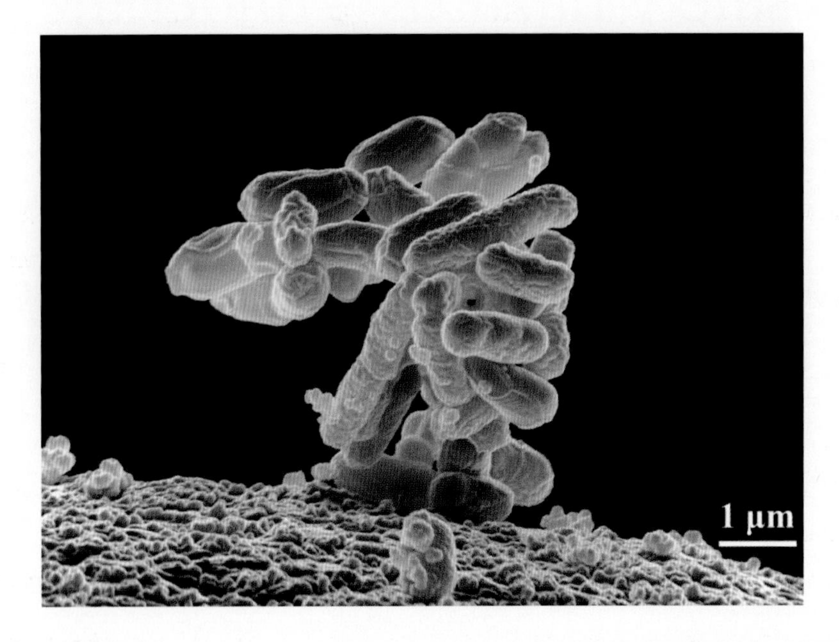

produce toxins that are harmful to humans. Further information on the toxins produced by moulds and algae is provided later in this chapter (under 'Chemical hazards').

Before reading further, it would be useful for you to learn a little about microorganism names, see Box 12.

Pathogenic foodborne bacteria

There are over a dozen different types of bacteria classed as foodborne pathogens. This book focuses on those that are more likely to be an issue for food businesses based in Australia.

Many bacteria exist in only one structural form, called vegetative cells. Some bacteria can also exist in another structural form called an endospore. For simplicity, throughout the remainder of this book, vegetative cells will be referred to as 'cells' and endospores as 'spores'.

Box 12 – Microorganism names

Like the scientific names for plants and animals (e.g. the domestic cat, *Felis catus*), the names of bacteria, fungi and parasites are written in italics and they are often derived from Latin words. The first part of these names is the 'genus' and the second part is the 'species'. For example, in the name *Felis catus*, '*Felis*' is the genus and '*catus*' is the species. There can be more than one species in a genus. For example, the Chinese mountain cat is *Felis bieti*. Two different bacterial species in the same genus are *Clostridium botulinum* and *Clostridium perfringens*.

It is common practice to shorten names of microorganisms after the first use by using only the first letter of the genus name. For example, *Clostridium botulinum* becomes *C. botulinum*.

You may be wondering how microorganisms get their names; here are a few examples:

- *Penicillium* was named after the Latin word for paintbrush, penicillus, because of how it looked under the microscope.
- *Escherichia coli* was discovered by a German doctor and scientist called Theodor Escherich and it can be found in the colon.
- *Salmonella* was named after Daniel Salmon, an American veterinary pathologist.

As with our cat example, some microorganisms are also known by non-scientific names. For instance, you may have heard the term 'Golden Staph' used for *Staphylococcus aureus*. This term is used because of the golden-yellow colour of some types of *S. aureus* when they grow.

Bacterial cells are simple structures, consisting of a cell wall enclosing the components that the cell needs to survive and grow, such as DNA. Bacteria need to be in their cellular form to grow.

Bacteria that form spores are termed 'spore-forming bacteria'. Spores have a thick outer layer that protects the inner contents from unfavourable conditions. For example, bacteria can survive exposure to chemicals and much higher cooking temperatures when they are in the spore form then when they are in the cell form. Although they are not capable of growing like cells, spores are able to survive for long periods of time in a dormant state. Spores can be activated to form cells again when the surrounding environment changes, such as when moisture or a new source of nutrients becomes available.

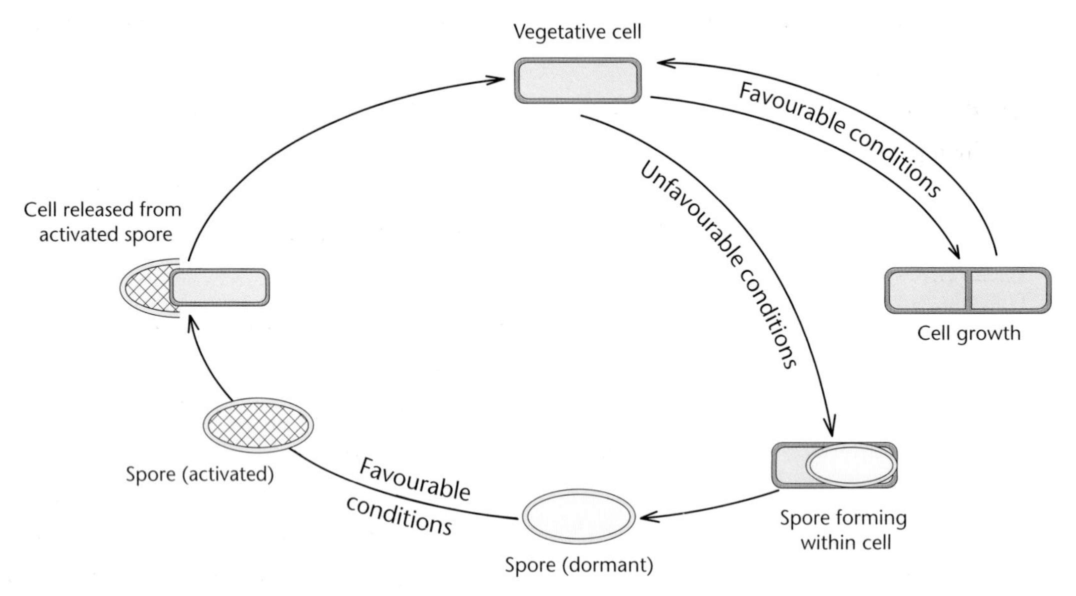

Bacterial cell–spore life cycle

How pathogenic bacteria can contaminate food

Pathogens can contaminate food either before or after it leaves the farm; that is, at any point in the 'food chain'. Routes or sources of contamination can be grouped into seven primary categories:

- **Food production animals (such as cows, pigs, sheep and poultry)** – pathogenic bacteria may occur naturally in the intestines of these animals and can be spread to the surfaces of their flesh (i.e. their 'meat') when the animals are slaughtered and butchered.
- **The environment where food is grown or harvested** – pathogenic bacteria that occur naturally in soil and water may be spread to food. For example, fruit and vegetable crops can become contaminated via the soil they are grown in, or fish and shellfish can become contaminated via the water they live in.
- **'On farm' contamination routes involving human practices** – there are many examples of these, including:
  - irrigating crops with water contaminated with animal manure
  - using contaminated water to dilute agricultural chemicals sprayed onto crops
  - using animal manure as a crop fertiliser
  - harvesting fallen fruit from orchards where animals have grazed
  - poor hygiene of field workers and lack of adequate toilet facilities

- spread of contamination from one area to another via contaminated farm equipment or contaminated water run-off.

See Box 13 for specific examples of foodborne illness outbreaks caused by these on-farm practices.

- **Insects and animal pests (post harvest or slaughter)** – pests that may spread contamination onto food include rats, mice, birds, cockroaches and flies. Contaminants can be spread by either direct contact or indirect transfer. An example of indirect transfer is people bringing bacteria into the premises by walking through an area fouled with bird droppings.
- **Food handlers** – people can carry pathogenic bacteria either as part of their normal body flora or because they have an infection, such as gastroenteritis. These pathogens may be transferred to food in a variety of ways, such as a food handler going to the toilet and not washing their hands properly before touching food.
- **Spread of contamination from one food item to another, or from contaminated equipment** – this is referred to as cross contamination (Box 14). Inadequate cleaning and sanitation, and unhygienic practices may spread pathogens in the food processing and handling environment; for example, via poorly cleaned equipment such as meat slicers, blenders and utensils.
- **Water used in food handling or processing activities** – use of untreated water during processing or failure of water treatment plants can contaminate food. However, these issues are not generally a problem in Australia.

Box 13 – Foodborne illness outbreaks caused by incorrect farming practices

- 1981 (Canada) – sheep manure contaminated with pathogenic bacteria was used to fertilise cabbage to be used in coleslaw: 41 people became ill (and 17 died as a result).
- 1996 (United States) – poor-quality apples and apples harvested from the ground were used to make unpasteurised apple juice: the juice was contaminated with pathogenic bacteria and 70 people became ill (and one died as a result).
- 1996 (United States/Canada) – water contaminated with pathogenic parasites was used to prepare pesticides sprayed on raspberries; bird droppings and infected field workers were also possible sources: 1 465 people become ill with gastroenteritis.

Box 14 – What is cross contamination?

Cross contamination is the spread of a contaminant from one food or piece of equipment to another. Contaminants that are of primary concern to Australian manufacturers are pathogenic bacteria, viruses and food allergens.

Examples of cross contamination include:

- pathogens in raw meat juices or soil on vegetables contacting ready-to-eat food
- food containing allergens is not cleaned off equipment properly and the allergens then contaminate a product that is supposed to be allergen-free.

The main ways of controlling cross contamination are:

- having an effective cleaning and sanitation program
- ensuring food is properly covered during storage
- storing raw and ready-to-eat foods apart.

Foodborne pathogenic bacteria – meet your enemies

The table below provides an overview of the different types of pathogenic bacteria that may contaminate products manufactured in Australia. More specific information about each type of pathogen is provided in Chapter 9. You may be particularly interested to learn more about those that are associated with the food types you handle.

Pathogenic bacteria and foods they have been associated with

Bacteria	Associated foods
Aeromonas species	Raw fish and shellfish, fresh produce exposed to untreated water
Bacillus cereus	Boiled or fried rice, porridge, pasta, processed meats, cooked vegetables, soups and sauces
Campylobacter species	Raw chicken, beef and offal (e.g. liver and kidneys)
Clostridium botulinum	Foods incorrectly preserved at home, smoked fish, vegetables in oil, incorrectly processed or cooled canned foods
Clostridium perfringens	Cooked meat, poultry, sauces, pies, casseroles and curries left to cool slowly at warm temperatures

Bacteria	Associated foods
Escherichia coli O157 and related types	Minced meat, salad vegetables, bean sprouts and sprouted seeds, and fermented smallgoods
Listeria monocytogenes	Coleslaw, soft cheeses, sliced processed meats, frankfurters, dips, pâté, cooked poultry and ready-to-eat seafood
Salmonella species	Poultry, raw or undercooked eggs, bean sprouts and sprouted seeds, and a wide range of fruits and vegetables
Shigella species	Ready-to-eat foods that are contaminated by an infected food handler who has poor hygiene or contaminated water used in food preparation
Staphylococcus aureus	Ham, cream-filled pastries, cheese and foods contaminated by a food handler
Vibrio cholerae	Raw seafood, fruits and vegetables washed in contaminated water or contaminated by a food handler
Vibrio parahaemolyticus	Raw fish, shellfish, and crustaceans (e.g. prawns, crabs, lobsters and crayfish)
Vibrio vulnificus	Raw oysters
Yersinia enterocolitica	Raw meat (particularly pork), raw poultry, unpasteurised milk and tofu

Because some foods or classes of food are at higher risk of causing foodborne illness owing to the presence of pathogenic bacteria, it is a requirement of the Code that they meet specific microbiological criteria (Standard 1.6.1 Microbiological Limits for Food). Products must pass these criteria at all stages of their manufacture or sale, until they reach the end of their shelf-life. Products that do not pass the criteria specified in the Code pose a risk to the health of consumers and therefore should not be offered for sale or further used in the preparation of food for sale (Standard 1.6.1).

Some foods are not allowed to have any detectable level of specific pathogens. For example, packaged cooked pâté is a ready-to-eat food that should have been heated sufficiently to kill bacterial cells. It is a requirement of the Code that when five 25 g samples of pâté from one batch or lot are tested, no *Listeria monocytogenes* or *Salmonella* must be found.

Foods listed in Standard 1.6.1 include:

- cheeses
- smallgoods

- prawns, crabs, lobsters and crayfish
- mussels, oysters, clams, octopus and squid
- sprouts (e.g. alfalfa, snow pea and mung bean).

It is your responsibility to check the Code to determine if the products you manufacture fall into any of these categories. If your products do need to meet specified limits, you must seek expert guidance because only a food microbiologist has the expertise to perform the tests required (see page 277). More information can be found in the User Guide to Standard 1.6.1 – Microbiological Limits for Food with additional guideline criteria (see page 270).

Growth of bacteria

Bacteria levels increase when cells make copies of themselves, in a process called 'reproduction' (commonly referred to in this context as 'growth'). One cell divides into two cells, each of which then divides into a further two cells giving a total of four cells, and so on. Bacteria grow when conditions in the environment surrounding them are favourable, and their number can double every 10 minutes. This means one single cell can become many thousands of cells in a short time. See the illustration opposite for a worst case example of fast cell growth when conditions are ideal.

Growth can stop for a variety of reasons. For example, as bacteria grow they use up the nutrients from their surrounding environment that are required to give them energy. If a new source of nutrients does not become available, growth slows down and eventually stops.

When you see pictures of bacteria that have been grown in a laboratory (see below), what you are actually looking at is cells clustered together in clumps called colonies. These colonies are only visible to the naked eye because they are each made-up of many thousands of individual cells that have grown from one single cell.

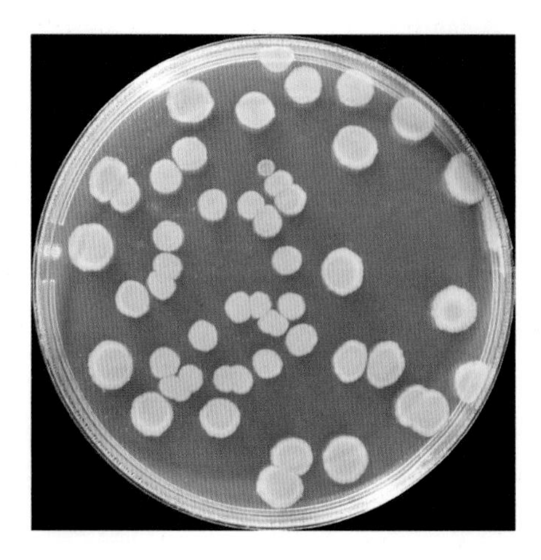

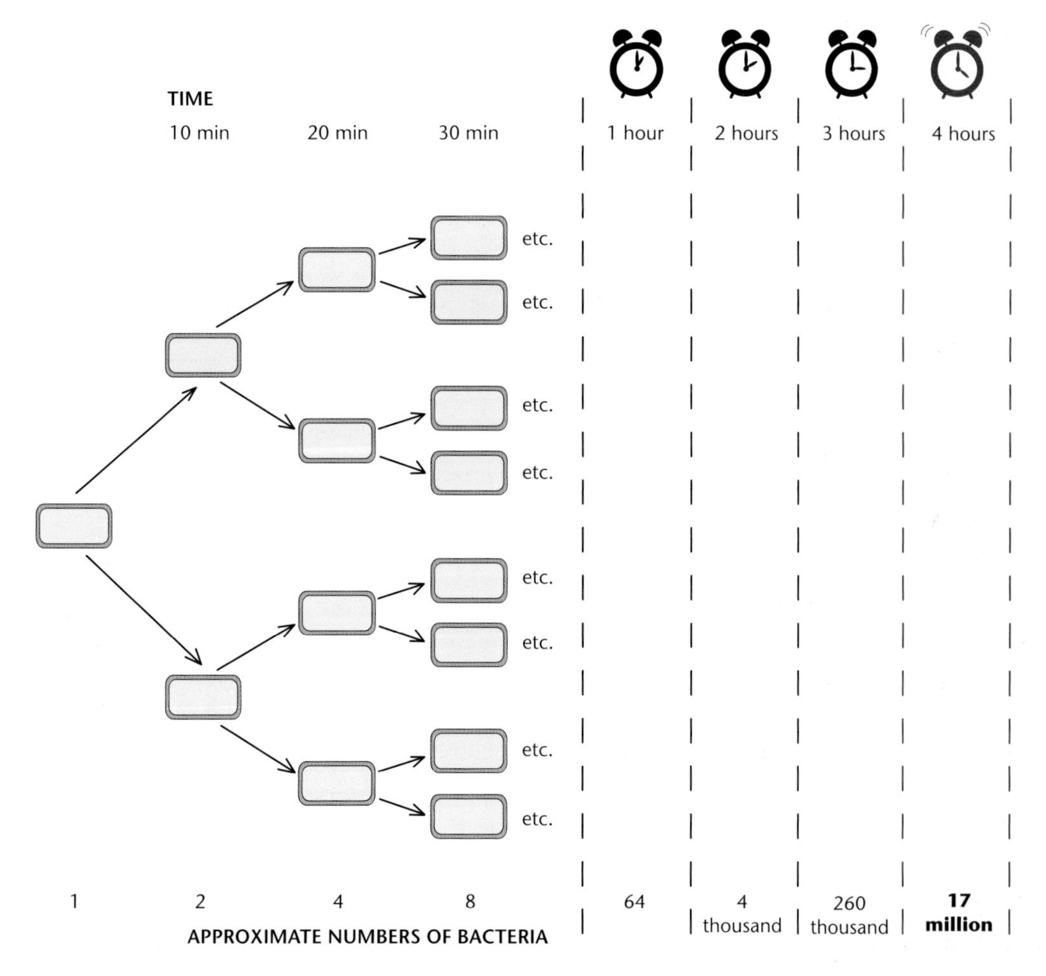

Rapid growth of bacteria under ideal conditions

Conditions bacteria need to grow in food or food handling areas

Like humans, bacteria and other microorganisms require certain environmental conditions to survive and grow:

- **Nutrients** – bacteria have much the same nutrient requirements for growth as people do. Once bacteria have used up the available nutrients, they stop growing.
- **Water** – bacteria require a certain amount of water or moisture to grow. Water not only needs to be present, it needs to be in a form that is accessible to the bacteria. Ingredients that bind up water and make it harder for bacteria to grow include salt and sugar.
- **Oxygen** – some bacteria need oxygen in their environment to grow, while others can only grow when there is no oxygen present. Some other bacteria can grow with or without available oxygen. Most foodborne pathogenic bacteria fit into this last group.

- **Temperature** – bacteria can survive across a wide range of temperatures, but can only grow in a specific, narrower temperature range. Within this temperature range, there is an even narrower range that is best for growth. For example, although *Listeria monocytogenes* is able to grow at refrigeration temperatures, it does so only very slowly. However, this still makes *L. monocytogenes* particularly hazardous for ready-to-eat foods that have a long refrigerated shelf-life (see Box 15).
- **Time** – once all of the above conditions have been met, bacteria still require time to grow. The time required to grow to a certain level is dependent on the bacterial species and the conditions in their surrounding environment.

It is possible for you to apply this knowledge – with a little more help provided in upcoming chapters – to your advantage by manipulating your product recipe or its storage conditions to make it an unfavourable place for pathogenic bacteria to grow.

Box 15 – *Listeria monocytogenes*: slowed-down but not stopped by a winter chill

Listeria monocytogenes is able to grow at refrigeration temperatures, even below 1°C. Chilling will, however, significantly slow its growth because its optimum growth temperature is 30–35°C.

Because *L. monocytogenes* can still grow at 5°C or below, chilling can't be relied upon as the sole food safety control for ready-to-eat foods that have a long shelf-life. This includes some types of soft cheese, some processed meats, cold-smoked fish and some pre-prepared salads.

L. monocytogenes is commonly found in food processing environments, particularly damp or moist areas (e.g. drains and evaporator trays in cool rooms). Following the recommendations provided in this book on hygiene, cleaning and sanitising practices will reduce the risks associated with this pathogen. Keeping ready-to-eat foods covered will also help.

It is extremely important that this pathogen is controlled in environments where ready-to-eat chilled foods are handled. It can cause serious illness and death in vulnerable populations. Pregnant women are of particular concern because *L. monocytogenes* can spread to the foetus causing miscarriage or stillbirth.

How pathogenic bacteria cause foodborne illness

Pathogenic bacteria can cause foodborne illness in two primary ways:

- Infection
  - After living pathogenic cells within food are eaten, they 'take-up residence' in the body and grow, which may or may not cause illness.
- Intoxication
  - Food containing toxins, produced by bacteria during growth in food, is eaten. The toxins, not the bacteria, then cause illness. Once the bacteria have produced the toxin, even if the food is cooked and the bacteria are killed, illness can still occur because some toxins are heat resistant and will not be destroyed by cooking.

Symptoms of infection usually take longer to appear than those of intoxication because there is generally a period when the bacteria need to grow to levels high enough to cause illness. This is called an incubation period. So, although people often think it was the last thing they ate that made them sick, it may in fact be something they ate 2 days earlier.

Pathogenic bacteria can be eaten without causing illness

It is possible for pathogenic bacteria to be eaten without causing any illness, for a variety of reasons:

- The pathogen may be present in the food in low levels, whereas relatively high levels are required to cause illness (unless eaten by vulnerable persons).
- The pathogen may be present in a form that is not able to cause illness, such as dormant spores.
- The pathogen may not be able to withstand the natural defence systems it encounters in our bodies, such as the acidic conditions of the stomach. This may be because the bacterial cells are not in a very healthy state or that generally they are not robust enough to survive unless protected in some way (see Box 86 in Chapter 9 for an example of this scenario).

The level of pathogenic bacteria required to cause illness varies

Relatively high levels of some types of bacteria have to be eaten to cause illness (e.g. a million cells per gram of food). Alternatively, they have to grow to relatively high levels before enough toxin is produced in the food to cause illness. Because it is very unlikely that high levels of any pathogen would initially be present in a food, this means that, for illness to occur, pathogens have to be able to grow in the food before it is eaten.

Unlike spoilage microorganisms, which give out 'signs' they are present (e.g. an off odour), most pathogenic bacteria can grow to high levels without causing any notable changes to how the food smells, looks or tastes. The highly contaminated food can be eaten without the consumer suspecting there is anything wrong, until symptoms of illness start.

However, some pathogenic bacteria can cause illness even if they are present at relatively low levels. Even fewer than 10 cells per gram of food may be enough to make someone sick. It is not always necessary for these pathogens to grow in the food for illness to occur; instead, the numbers initially present in the food may be enough.

Primary control measures for pathogenic bacteria

As previously defined, control measures are the actions that can be taken to eliminate hazards or to reduce the risks associated with them. The focus here is on actions that can be applied in food processing and handling settings:

- **Prevention** – one of the key control measures is to prevent pathogens from contaminating food during storage, handling and processing. One way to achieve this is to have appropriate packaging for heat processed foods, to prevent pathogens from re-contaminating food after processing.
- **Heating** – relevant pathogens are killed by heating food to a minimum temperature during cooking and holding at this temperature for a minimum time. This is called a time–temperature combination. The spores of pathogenic spore-forming bacteria require more severe time–temperature combinations to control than bacterial cells (Box 16).
- **Chilling** – cooling foods to 5°C or below prevents or considerably slows the growth of pathogenic bacteria. See Box 17 for the reason why cooked foods must be chilled down rapidly. As some pathogens can grow slowly at refrigeration temperatures, potentially hazardous foods can be stored only for a limited time.
- **Checking and/or adjusting the recipe** – some foods do not naturally support the growth of pathogenic bacteria, such as lemons because of their acidity. Other foods can be adjusted to slow or stop the growth of pathogens by, for example, adding lemon juice to a food that is naturally low in acid.

It is also common practice to combine several control measures. One example is heating food using a specific time–temperature combination then chilling it to 5°C or below. By itself, each control measure may not completely kill or stop the growth of pathogens, but by adding two or more together they can become effective overall. The term given to this method of controlling microorganisms is the 'hurdle concept'; see Box 18 for an explanation.

Box 16 – Pathogenic spore-forming bacteria

Clostridium botulinum, Clostridium perfringens and *Bacillus cereus* are pathogenic spore-forming bacteria. They can be present at low levels in many foods that are grown in soil or that may come into contact with soil; in other words, in most raw commodities. These spores are usually dormant and only form cells when environmental conditions become favourable.

Standard cooking practices are unable to kill these heat resistant spores; they are the toughest form of bacteria to kill. Cooking may actually activate the spores from their dormant state so they can form cells and grow. Specialised heat processing techniques, such as those used to produce canned foods, are the only way to effectively kill most of these spores.

Pathogenic spore-forming bacteria can also be controlled by preventing or slowing their growth. This can be achieved by storing products at 5°C or below for a limited time and/or adjusting product recipes (e.g. adding acid).

Box 17 – Rapid chilling: why is it necessary?

Clostridium perfringens spores are able to survive standard cooking processes. In fact, heating can activate dormant spores, causing them to change into cells, which can then grow.

At its optimum growth temperature of 44°C, *C. perfringens* can grow much more rapidly than other pathogenic bacteria. Cell numbers can double every 10 minutes, so if only 10 cells per gram of food were initially present, they could multiply to high enough levels to cause illness in just over 3 hours. This is why it is very important to stop the bacteria from growing by cooling food to 5°C or below as quickly as possible.

The Code specifies time–temperature requirements that must be followed to cool down heat processed foods to 5°C or below. Following these will not allow *C. perfringens* to grow to high enough levels to cause illness. See Chapter 6 for details.

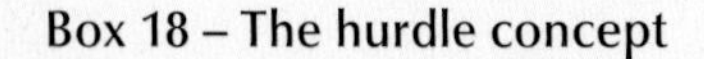

Box 18 – The hurdle concept

In this example of the hurdle concept, the bacteria faced four hurdles (i.e. control measures): heat processing, addition of acid, addition of salt and storing at 5°C or below. As you can see, in this case all four hurdles needed to act in combination before all the bacteria were finally 'defeated'.

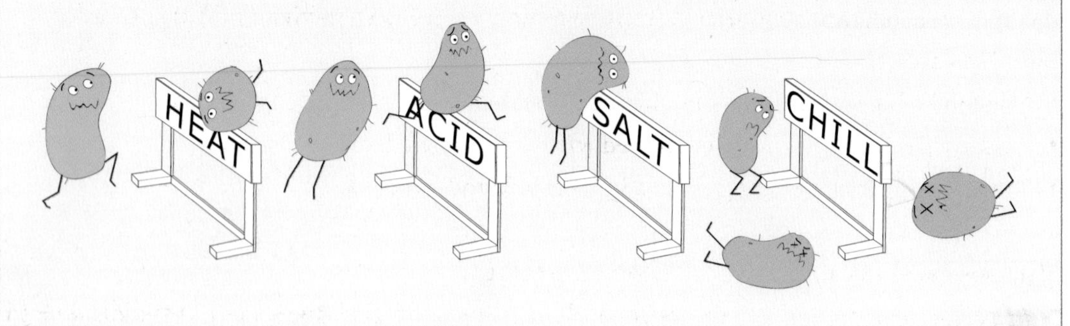

Pathogenic foodborne viruses

There are approximately 14 different viruses that are classed as foodborne viruses. In this book, the focus will be on those that are of primary concern for the Australian food industry: hepatitis A, noroviruses, astroviruses and rotaviruses.

Viruses are simple structures consisting of genetic material surrounded by a layer of protein. They are so simple that they are not actually cells and are called 'particles'.

Unlike other microorganisms, viruses are not able to grow or multiply by themselves. To multiply, viruses have to invade living cells, which then become host cells. Once inside host cells, viruses are able to make multiple copies of themselves. Eventually the host cells burst and release the new virus particles, which can then invade other host cells.

Although viruses cannot grow in food, only very low levels need to be eaten to cause illness.

How pathogenic viruses can contaminate food

Food can either be contaminated with viruses at the point where it is grown or harvested, or during 'post-farm' handling. Viruses are primarily spread by humans, unlike many bacterial pathogens, which are often associated with animals used in food production. Routes of viral contamination include human sewage contamination of waterways and water supplies (e.g. crop irrigation water) or poor hygienic practices used by infected food handlers (e.g. not washing hands after going to the toilet).

Oysters grown in water contaminated with sewage have caused large-scale viral foodborne illness outbreaks. Because oysters filter large volumes of water during feeding, they can accumulate high levels of virus particles. Apart from oysters, viruses do not tend to be associated with specific foods; rather they can occur on any food from a contaminated environment or handled by a contaminated person. Foods commonly implicated in viral foodborne illness outbreaks are:

- seafood – harvested from contaminated waterways
- salad vegetables – irrigated or washed with contaminated water
- ready-to-eat foods – contaminated by infected food handlers.

Viruses can have extended incubation periods

The incubation periods for noro-, astro- and rotaviruses are much the same as infections caused by pathogenic bacteria. Hepatitis A, however, can take up to a month (even 6 weeks) to 'make itself known' to its host. This can cause problems for food handlers and their employers, and officials who are investigating foodborne illness outbreaks.

Primary control measures for pathogenic viruses

There are fewer food safety control measures for viruses than for bacteria. This is because measures associated with preventing growth are not relevant because viruses can't grow in food.

These are the primary control measures for pathogenic viruses in food processing environments:

- Ingredients should be purchased from **reputable suppliers**.
- **The work attendance or duties of food handlers should be restricted if they show symptoms of foodborne illness**; staff diagnosed with a foodborne illness must not perform any food handling duties until a doctor advises they are fit to do so again.
- **The work attendance or duties of food handlers who have been in contact with a person infected with hepatitis A** should be restricted until a doctor has given them clearance.
- **Hygienic practices should be used at all times**, particularly as food handlers may not even know they are infected with a virus, such as during the long incubation period for hepatitis A. These practices must be even more rigorous when handling ready-to-eat foods.
- **Food** should be heated using the same time–temperature combinations used to control bacterial cells. Heating to 70°C for 2 minutes destroys most types of pathogenic foodborne viruses.

Parasites

Parasites are more complex microorganisms than bacteria and viruses. They are able to survive and replicate by getting their food from other living creatures, which are termed hosts. During their life cycle, parasites can move between multiple hosts or environments (see Box 19 for an example). Although parasites may cause illness in their hosts, it is not in their interest to kill them because this limits their supply of food, so some parasites can live within a host for extended periods, such as intestinal worms in a dog.

Parasitic microorganisms belong to two major groups – protozoa (e.g. *Giardia*) and helminths (e.g. tapeworms). Numerous types of parasites cause significant food safety issues in developing countries, but only those that are more relevant to Australia, *Giardia duodenalis* and *Cryptosporidium parvum*, will be discussed in any detail here.

However, it is important to note that many of the parasites that are not discussed here are commonly associated with animals (wild, domestic or food production). It is essential that animals are kept well away from food preparation areas and that food handlers who come into contact with animals follow strict personal hygiene practices.

Protozoan parasites – *Giardia* and *Cryptosporidium*

Giardia and *Cryptosporidia* are major causes of diarrhoeal illness worldwide, but most notably in areas where there are poor hygiene and sanitation standards. Both parasites can infect a wide variety of animal species. *Giardia* hosts include dogs, cats and numerous wild animals, including Australian marsupials. *Cryptosporidia* can be found in reptiles, birds and mammals.

Salad vegetables, fresh herbs and ready-to-eat foods are most often implicated in foodborne illness outbreaks caused by these parasites. Unpasteurised goats milk and unpasteurised apple juice have also been implicated.

Watery diarrhoea and abdominal cramps are the primary symptoms of gastroenteritis caused by *Giardia* and *Cryptosporidia*. People with weakened immune systems may die from the illness.

Both parasites can exist in two forms. Oocysts are the dormant form that can survive for extended periods in the environment. Oocysts are more resistant than the other form to factors such as chemical sanitisers or drying. The second form is a vegetative one, which is capable of replication inside the host. The vegetative form moves through the host's digestive system, replicating and then converting to oocysts, which are then shed in high numbers in faeces. This shedding can continue for weeks or months after symptoms have ceased. This pattern means

Box 19 – Parasite life cycle

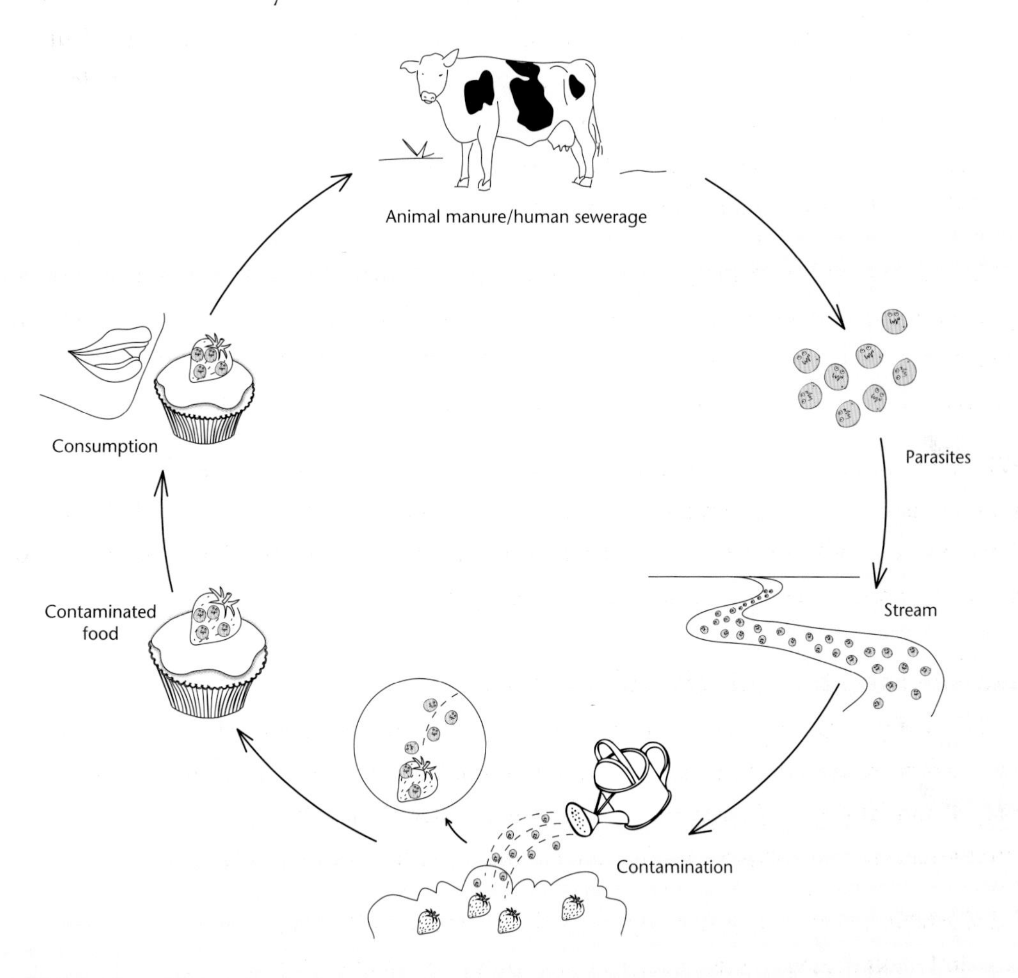

During their life cycle, parasites can move between multiple hosts or environments. There are two primary ways parasites can be transferred to food:

- contamination by animal manure or human sewage of:
  - crop irrigation water
  - water used for washing fresh produce
  - water used as an ingredient in food.
- transfer by food handlers who are infected with the parasite and shed it in faeces.

Parasites are not able to grow or replicate in water or food. However, only very low levels generally need to be eaten to cause illness.

that food handlers who are infected pose a real risk of spreading their infection if handling ready-to-eat foods. In addition, food handlers can re-infect themselves after their symptoms have gone if they don't practise proper hand washing.

The control measures for these parasites are similar to those required for foodborne pathogenic viruses:

- **heating** food using standard cooking practices
- food handlers using a high level of **personal hygiene**, especially when washing hands
- **restricting work attendance or duties for those suffering symptoms of gastroenteritis** until recovered.

Food handlers who suffer from numerous bouts of diarrhoea must be placed on alternate duties until a medical diagnosis is provided. Then they must only be allowed to recommence normal duties when they have received the 'all clear' from a doctor.

Chemicals – environmental contaminants, food business use or naturally occurring

The following is an overview and some examples of chemical food safety hazards. Many of these are beyond the control of small food business owners. It is useful, however, to learn something about them so you can understand how effectively these issues are controlled in Australia, ensuring they do not cause significant levels of illness or affect our export market.

Government regulations and monitoring

FSANZ, AQIS and other Australian government authorities strictly regulate and monitor potential chemical food safety issues associated with:

- agricultural and veterinary practices
- environmental contamination
- toxin-producing microorganisms
- food packaging materials
- misuse of food additives.

This is done in a variety of ways, including specifying and monitoring limits for chemical contaminants and natural toxins in nominated foods (e.g. mercury in fish). Maximum residue limits (MRL) are set for agricultural and veterinary chemicals and are incorporated in the Code.

An MRL is the highest level of a chemical allowed to be present in a food. The levels are specified for pesticides, herbicides and chemicals used by vets to treat animals. MRLs are mainly set for fresh produce and animal products, not for processed foods. If the level of chemical is found to be higher than the MRL, it is likely the chemical has been used improperly or incorrectly; for example, farmers not waiting long enough between spraying crops with pesticide and harvesting.

Australian government bodies regularly collect and analyse samples from farms, retail outlets and some imported foods to ensure that levels of introduced chemicals and natural toxins are within specified limits. In Australian foods, levels are usually found to be within the set limits. However, if you use imported ingredients, it may be worthwhile asking your supplier to provide a certificate of analysis, at least for the first shipment, to demonstrate that any agricultural chemicals used on the product are permitted in Australia and that they are within limits. Checks should also be made to ensure that permitted levels of natural toxins are not exceeded. If the supplier sources the raw material from a new producer, this process might need to be repeated.

Chemicals hazards in food businesses

There are many different chemicals used in food businesses. Due to the care taken when handling these chemicals in Australian food businesses, cases of health issues associated with them are very rare here. However, there have been food recalls in Australia owing to accidental contamination or misuse of approved additives.

Food additives

In Australia, all food additives undergo rigorous safety assessments before being listed in the Code as an approved substance. The Code specifies the chemicals permitted for use and also which classes of food they are permitted in and at what level. You can find this information in the following standards:

- 1.3.1 Food Additives – for example: preservatives, artificial sweeteners, acids, food gums, colours and flavouring agents.
- 1.3.2 Vitamins and Minerals – this does not include substances naturally present in foods – only those that are added separately.
- 1.3.3 Processing Aids – for example: lubricants, bleaching and washing agents, filtering and clarifying agents, and antifoam agents.

Certain substances are more of a food safety issue than others. If you add vitamins or minerals to foods, you need to be aware that some can be toxic if the levels are too high (e.g. iron and Vitamin A). Nitrite preservatives are also toxic at high levels and may cause death within minutes of being eaten because the body's ability to transport oxygen in the bloodstream is affected (see 'Adding chemical preservatives' in Chapter 4 for further information). Some people are sensitive to sulphite preservatives and the presence of more than 10 mg/kg of food must be stated as a warning on product labels.

Specific control measures for safe use of chemicals include:

- **Only use substances permitted in the Code** for your specific product types – seek advice from a technical expert, such as a product development consultant or your local food authority if you have difficulty finding the information in the Code.
- **Do not exceed the permitted levels specified in the Code** – seek advice if you are unsure how to calculate appropriate concentrations for your products (never guess); ensure that only staff that have received specific training handle these substances; weigh out the chemicals separately from other ingredients following prepared recipe sheets.
- Ensure you **store chemicals safely** – keep them separate from food ingredients and ensure storage containers are clearly labelled.
- **Provide your staff with training** on how to correctly weigh out or measure chemicals, and ensure that weighing equipment is maintained according to the manufacturer's instruction.

Chemicals used for cleaning and sanitising

Numerous substances are used as cleaning and sanitising agents in food preparation settings. The key to making sure they do not become a food safety hazard is to carefully follow the instructions provided with the products, particularly those regarding the need to rinse off after use. It is also important to keep the chemicals adequately separated from food ingredients in well-labelled containers in a dedicated storage cupboard. Use of dedicated scoops and containers is essential if the chemicals need to be diluted before use. Never store chemicals in a container that looks the same as your food storage containers. More information on the safe use of chemicals for cleaning and sanitising is provided in Chapter 3.

Pest-control chemicals

Just like chemical sanitisers, there are a wide range of chemical pest-control substances available in a variety of forms (e.g. sprays, pellets or powders). Many of these are toxic to humans.

The best way to stop these from accidentally contaminating foods is to use professional pest-control contractors that have appropriate accreditation or credentials.

If you do choose to 'do-it-yourself', here are some control measures to follow (see Chapter 3 for more details):

- Have a documented **pest-control program**.
- **Store pesticides safely** – separately and preferably locked away from food ingredients.
- **Store pesticides in original containers**, but if you do transfer them, clearly label the new container and make sure it does not look the same as any food storage containers.
- Be aware that sprays will spread great distances, so **only use sprays after covering up or putting away all food and utensils**.
- **Make sure that instructions for use of specific products are read and understood by staff** before they use the product for the first time.

Equipment maintenance

Greases and lubricants used for equipment and vehicle maintenance must also be handled carefully. Follow the same storage recommendations as for other types of chemicals. Make sure greases or lubricants used on food processing equipment are suitable for food contact (i.e. they are not toxic). It is also a good idea to check if greases and lubricants contain any food allergens, such as oil made from fish, peanuts or soy.

Regularly inspecting the inside of stainless steel equipment for signs of corrosion, which may cause metal contaminants to leach into food, is also important. Leaching is more likely to occur if you use corrosive sanitisers incorrectly or process corrosive foods, such as tomato products or foods containing sulphur dioxide.

Packaging materials

Packaging can be another source of chemical contamination and only 'food grade' packaging should be used for food. Packaging should also only be used in the manner in which it was intended to be used. For example, heating of some types of plastics can cause the release of chemicals that are thought to be toxic. Only packaging that is classed as suitable for heating should be used for products that are hot filled or heated in the pack by the manufacturer or the consumer, and then it should only be heated to the temperature specified by the packaging company. In general, recycled packaging should not be put in direct contact with food, because ink and other residues from the recycling process may transfer into the food, especially food containing fat.

Naturally occurring substances

Chemical toxins can be naturally present in foods, and specific food types contain substances to which individual people are allergic or sensitive. In general, these toxins and substances are able to resist standard cooking practices and they can't be detected by consumers (i.e. they have no taste or odour).

Food allergens

Food allergens are naturally occurring proteins. When eaten, these can cause abnormal immune responses in sensitive individuals. Allergic reactions vary from mild illness (e.g. skin rashes) through to severe, life-threatening anaphylactic shock involving swelling of the airways, possibly leading to death.

Illness caused by substances in food can also be food intolerances or sensitivities, rather than allergic reactions. However, as the food safety controls for substances causing allergic reactions are identical to those used for substances causing intolerance reactions, they are grouped together as one here and are collectively called 'food allergens'.

Food allergens of primary concern in Australia are:

- cereals containing gluten and their products; namely wheat, rye, barley, oats and spelt, and triticale (a cross between wheat and rye)

- crustaceans (e.g. prawns, crabs, lobsters and crayfish) and their products

- egg and egg products

- fish and fish products

- milk and milk products

- peanuts and soybeans and their products

- sulphites, when added in concentrations of 10 mg/kg of food or more

- tree nuts (almonds, Brazil nuts, cashews, hazelnuts, macadamias, pecans, pine nuts, pistachios, walnuts) and sesame seeds and their products.

Note – coconut is not classified as a nut, and is not considered to be a food allergen.

All the food allergens listed above are subject to specific labelling requirements in the Code and these are discussed in Chapter 7. The only way consumers can manage their food allergy or intolerance is to avoid eating foods containing the substance or substances to which they are allergic or sensitive. Therefore, it is a key requirement that product labelling is correct. This means you must be certain which products contain, and do not contain, food allergens.

Allergens may be present in food products because they are a natural component of a whole food or are present in an ingredient. They may also be introduced via cross contamination. This type of cross contamination is called cross contact. Allergens can contaminate food through use of equipment and utensils that were in previous contact with allergen-containing products. A tool called Voluntary Incidental Trace Allergen Labelling (VITAL) has been developed to assist food businesses to assess the risk of cross contact throughout all stages of product preparation (Box 20).

The primary control measures for keeping you products safe for consumers with food allergies are:

- **Have a sound understanding of which raw materials you use are food allergens or contain food allergens**. This requires knowing exactly what is in every ingredient you use (e.g. does the mayonnaise contain egg?). If you purchase multi-component raw materials, you need to know which individual ingredients they contain. Remember to check spices, flavours, colours and any processing aids (e.g. enzymes).

- **Consider the possibility that even ingredients that are classed as 'allergen-free' may have been cross-contact contaminated with allergens** before you received them and you must determine if this is the case. For information on how you can investigate this, see Chapter 5 (under 'Purchasing your ingredients').
- Change the ingredients used to **reduce the number of food allergens you handle** (where possible).
- **Know how to correctly label products** that contain, or may contain, food allergens, and make sure product labels are always correctly matched to products.
- **Minimise the chance of cross-contact contamination occurring**:
  - use dedicated product lines where possible
  - if you don't use dedicated product lines, schedule the production of allergen-free products before allergen-containing products
  - have a robust cleaning and sanitation program (Chapter 3)
  - store food allergens in well-sealed and labelled containers
  - provide staff with adequate training and make sure they know the importance of minimising the risk of cross contact
  - have strict control over any re-work (Box 21).

Box 20 – Voluntary Incidental Trace Allergen Labelling (VITAL)

VITAL is a risk-assessment tool that allows food processors to assess the potential for food allergen cross contact during all phases of product preparation. This assists food business owners to determine if precautionary allergen labelling is required on their products. It also assists businesses to set up their ongoing monitoring programs so any changes to the level of risk are detected and responded to promptly.

The VITAL tool was developed to reduce the indiscriminate use of precautionary labelling. This is when products are labelled with statements such as 'Produced on the same equipment as …' without an assessment for the actual potential for cross-contact contamination occurring.

VITAL will guide you through the following steps:

- assessing the allergen status of raw ingredients used
- reviewing the production line used to manufacture products

- calculating the amount of an allergen that may be present in products (through cross contact)
- using the calculation results to decide if precautionary labelling is required or not.

The VITAL tool is available free of charge on the Allergen Bureau website; for details see page 277. There are also worked examples and a 'question and answer' section available on this website.

Further information on labelling of products that contain, or may contain, food allergens is provided in Chapter 7.

Box 21 – Re-work and food safety

The term re-work is used to describe the process of using leftover ingredients or products from one production run in another production run (of the same or a different product).

Examples include using:
- leftover cooked meat as a pizza topping
- leftover dough for baked goods
- broken or poorly formed vegetable patties in a vegetable pastie
- leftover ingredients to prepare a cooking stock or soup
- stale bread as breadcrumbs.

There is a risk that allergens in leftovers will accidentally be incorporated into products that are intended to be allergen-free. There is also an increased chance that there will be higher levels of pathogenic bacteria in leftovers because they may have been left out at room temperature for longer than fresh ingredients.

It is recommended that any leftovers are clearly labelled with a full list of their ingredients and the presence of any food allergens is highlighted. Staff should be instructed to always talk to their supervisor before using any leftovers as re-work.

If re-work materials are regularly used, it is recommended that written procedures for handling these are developed. These should state the specific types of leftovers and in which products they may be used. In general, these are only those with the same allergen types. Prohibit unauthorised use of any leftovers not covered by these procedures.

Mycotoxins

Mycotoxins are chemical compounds produced by some types of moulds. Fortunately, the moulds that commonly spoil manufactured foods (e.g. mould growth on bread) do not usually produce mycotoxins. However, certain agricultural commodities may be contaminated with mycotoxins, either before or after harvesting. The following table gives some examples.

Primary mycotoxins of concern to the international food industry

Mycotoxin	Of main concern in	Potential health effect
Patulin	Fruit juices (particularly apple)	Genetic mutations
Ochratoxin A	Cereals, wine, coffee and grapes	Kidney disease
Zearalenone	Cereals	Endocrine system disruption
Aflatoxins	Peanuts, tree nuts and maize	Liver cancer
Trichothecenes	Cereals	Acute vomiting, diarrhoea (if large amount eaten)
Fumonisins	Maize	Kidney and liver disease

There is no evidence that mycotoxins cause any notable health problems in Australia. In developing countries, however, long-term consumption of contaminated cereal grains, nuts and fruit has been linked to illness.

As harvest conditions in Australia are mainly warm and dry, the threat of mycotoxin production in agricultural commodities is much lower compared with other regions (particularly tropical regions). However, if you intend to source any of the foods in the above table as organically grown, you should request to see regular reports on mycotoxin level checks.

Contamination of Australian peanut crops by aflatoxins, however, is an ongoing issue. Peanuts sold at the commercial and wholesale level in Australia do not exceed the permitted mycotoxin levels set by Australian authorities because this hazard is controlled at the growth, harvesting and packing level (e.g. automatic screening-out of mouldy nuts).

Food businesses that use peanuts and tree nuts as ingredients should check that their suppliers are aware of the potential issue of mycotoxins and can provide a guarantee that specified limits are not exceeded (see Chapter 5 for further information).

Marine toxins

Toxins produced by algae or bacteria may be present naturally in fish and shellfish.

Algal toxins can contaminate fish and shellfish directly (when they eat toxic algae) or indirectly (when they eat smaller fish that have been contaminated with algae toxins). Over time, the toxins can build up to high levels in seafood because they do not adversely affect fish or shellfish and are not readily excreted from their systems.

Ciguatera poisoning is an example of illness caused by eating fish contaminated with algal toxins. Symptoms include numbness around the mouth, nausea, vomiting, diarrhoea, dizziness and blurred vision.

There are four illnesses that may be caused by eating shellfish contaminated with algal toxins, these are:

- paralytic shellfish poisoning, which can be life-threatening if the muscles required for breathing are severely affected
- diarrhetic shellfish poisoning, which causes severe gastroenteritis
- neurotoxic shellfish poisoning, which causes gastroenteritis and prickling or burning sensations around the mouth
- amnesic shellfish poisoning, which can lead to permanent brain damage and, in some cases, death.

Australian authorities run water-monitoring programs in fish and shellfish harvesting areas that have had past issues with toxic algae species. If there is particularly heavy algae growth (called algal blooms), authorities will notify commercial and domestic fishers and advise them to cease catching or harvesting in the area until testing confirms that safe levels have returned.

Scombroid poisoning is caused by high levels of chemicals called histamines produced by spoilage bacteria in certain fish. The levels may become high enough to cause illness even if the fish appears to be of acceptable quality. Symptoms of scombroid poisoning are very similar to a severe allergic reaction and include burning or tingling around the mouth, rash or swelling over the chest and neck, itchy skin, nausea and vomiting. The primary control measure is to prevent or slow the growth of spoilage bacteria by chilling fish to 5°C or below as quickly as possible post catching, and minimising the amount of time fish is stored or handled above 5°C.

Poisonous plants and fungi

The Code lists nearly 200 plants and fungi that are prohibited from being used in food because of their potential to cause illness or death (Standard 1.4.4 Prohibited and Restricted Plants and Fungi). However, the vast majority of these are not commonly thought of as a food source. In Australia, naturally toxic plants are generally only an issue for farmers whose livestock graze on these species (many of which are native plants).

Food businesses in Australia do need to be aware of toxic mushrooms, toxins that may be found in potatoes, and the need to inactivate a toxin found in red kidney beans.

Many different poisonous mushrooms grow in Australia, including the most dangerous of all, the 'Death Cap'. The primary control measures are to purchase mushrooms only from reputable suppliers and never collect your own mushrooms unless you have received detailed guidance from an expert.

Potato plants are able to produce their own protective chemicals called glycoalkaloids, which are lethal to insects, animals and microorganisms. Glycoalkaloids may be present in potatoes that have been exposed to light, bruised or damaged. These chemicals cause gastroenteritis, and even death. The presence of glycoalkaloids in the potato is often, but not always, indicated by a green colour. Storing potatoes in the dark and avoiding use of green potatoes are the primary control measures.

Raw or semi-cooked red kidney beans contain a compound called phytohaemagglutinin, which is toxic to humans. The symptoms of the illness caused by this toxin are severe nausea, vomiting and diarrhoea, and usually last for 3–4 hours. Eating only four or five raw beans can be enough to cause illness. Notably, semi-cooked kidney beans may be more toxic than raw beans. To inactivate the toxin, pre-soaked beans should be boiled in fresh water for at least 10 minutes.

Physical contaminants in food

Physical contaminants in food – also called foreign bodies or foreign objects – rarely cause any severe harm to consumers. They are generally either seen before being eaten or they are soft or small enough to be harmless. However, hundreds and thousands of dollars are spent by Australian food businesses every year to recall products containing physical contaminants. The reputation of food businesses can be severely damaged by these occurrences. The presence of physical contaminants can bring on a strong emotional response in consumers, who then may regard the overall food handling practices of the responsible business as 'suspect'.

Physical contaminants can be split into two groups. The first are contaminants that are present because they were part of the original raw ingredients; these include bones, woody plant material, seeds and pips. The second group consists of items such as insects, wood, glass, metal or plastic, which may contaminate either raw or processed ingredients.

There are three main ways that physical contaminants can become present in food products:

- **Inadequate removal or contamination during primary production,** including harvesting, sorting, processing and transport; examples include:

  - bones not fully removed from fish or meat
  - stones and insects gathered up with fresh produce during harvesting
  - metal from farm machinery contaminating fresh produce during harvesting
  - inedible parts of fresh produce not completely removed during sorting and packing
  - lead shot from shooting game
  - packaging material parts, such as nails from pallets.
- **Introduced during processing and handling by food businesses,** examples include:
  - personal items of food handlers (e.g. jewellery or pens)
  - food handler wound dressings
  - pieces of poorly maintained equipment or fittings (e.g. shards of rust, or nuts and bolts)
  - pieces of packaging materials (e.g. broken glass or plastic caps)
  - insects and pests (e.g. cockroaches or bird feathers)
  - tools and miscellaneous items (e.g. paper clips, staples or cable ties).
- **Intentional** – deliberate contamination, which may occur either at the food business or at the place of sale.

The primary control measures to minimise the risk of physical contamination of food products are:

- **Prevention** – take steps to eliminate potential sources and risky handling practices; for example:
  - check that product specifications for raw ingredients cover physical hazards
  - keep a record of physical contamination of raw ingredients and change suppliers if it becomes a frequent issue
  - have an adequate pest-control program
  - prohibit staff from wearing jewellery or taking personal items into food handling areas
  - monitor the use of wound dressings on hands, which should be protected by a waterproof dressing or disposable gloves
  - avoid purchasing equipment that may become a physical hazard; for example, do not use glass thermometers in food handling areas
  - modify equipment, fixtures and fittings; for example, buy plastic covers for glass light fittings or redesign equipment to reduce the chance of parts falling into food
  - have a maintenance schedule for equipment; for example, checking for loose, perished or rusted parts that can fall into products or torn sieves that can let foreign bodies through
  - look after, and regularly replace, tools and other items used for equipment maintenance so they do not themselves become a physical contaminant
  - do not store non-food items in food packaging.

- **Detection** – use manual or automated systems to detect physical contaminants in ingredients or products; for example:
  - educate your staff on the importance of being alert for the presence of physical hazards
  - where necessary, use sorting or inspection lines dedicated to separating out contaminants from ingredients (e.g. inspectors or vibrators to remove small stones)
  - rotate staff on inspection duty frequently to make sure they remain alert
  - use automated detection equipment, such as a metal detector, if appropriate.
- **Removal** – either routine removal or in response to detection of a specific contaminant; for example:
  - routine washing of fresh produce to remove soil, insects and plant material
  - routine sieving of powders to remove any objects
  - have pre-printed 'Reject – not safe' labels ready for contaminated products that may need to be separated and disposed of
  - have dedicated disposal bins or hold areas for rejected items
  - have a glass breakage procedure (see Chapter 7, Box 66).
- **Investigate and implement corrective actions** – determine the source of the contaminant and take steps to prevent future occurrence. You must also identify the product batches that may be affected and must be recalled.

Although much emphasis is placed on avoiding use of glass in food handling areas, it is worth pointing out that clear brittle plastics can be just as hazardous.

KEY MESSAGES FROM CHAPTER 2

- A foodborne pathogen is a microorganism that can cause illness by being present in food or producing a toxin in food.
- Pathogenic bacteria and viruses are the microorganisms most likely to cause isolated cases or outbreaks of foodborne illness in Australia.
- Some pathogens only need to be present in food at very low levels to cause illness when eaten.
- The risks from pathogens in food processing environments can be reduced by:
  - preventing the pathogens from contaminating food
  - heating food to kill or reduce levels
  - chilling foods to stop or slow growth
  - adjusting recipes to stop or slow growth
  - using the above factors in combination.
- Standard cooking practices are unable to kill heat resistant bacterial spores; rapid cooling of food after cooking is a key control measure for spore-forming bacteria.
- Chemicals must be handled with care to stop them becoming a food safety issue; these include food additives, cleaners and sanitisers, and pest-control chemicals.
- Food allergens can cause life-threatening illness. To control food allergens:
  - correctly and clearly label products they are in
  - minimise the risk of them contaminating allergen-free foods.
- Physical contaminants include bones, woody plant material, seeds, insects, wood, glass, metal or plastic.
- The risks from physical contaminants can be reduced by:
  - taking steps to prevent contamination
  - monitoring food for their presence
  - using manual or automated systems to remove from food
  - investigating the cause of any contamination and taking steps to prevent re-occurrence.

Chapter 3

Controlling food safety hazards – premises and people

Now you are aware of the different types of food hazards, you need to know how to reduce or eliminate the risk of these becoming a problem in your food business. Let's start with the very basics – the safety of the environment in which your products are prepared and the actions of the people preparing them.

Food business premises and equipment

Poorly maintained or dirty premises can provide hiding places for foodborne pathogens that can then contaminate your products. Food allergens can also contaminate allergen-free products because of these issues. Neglected equipment can result in under-processed products or it may break down and cause physical contamination of your products.

Safe products cannot be made in an unsafe environment

Ideally, food products should be produced in purpose-built facilities, but this is not always possible, particularly for small businesses just starting out. If you have been given approval from your local council or relevant authority, you may begin operating out of your own home using a standard domestic kitchen. Later you might be able to scale up and lease premises that may or may not be designed for food handling. Eventually, if the business is successful, a purpose-built factory may be required. Food may also be manufactured or processed on a small scale using multi-purpose premises such as a restaurant kitchen or the back of a fruit and vegetable shop or delicatessen.

Regardless of the location, there are fundamental requirements you must meet for the safety of your products. These are:

- the layout of the premises minimises opportunities for food contamination
- the design of the premises and equipment allows effective cleaning and sanitising
- essential services and equipment are available, such as hand washing facilities, a source of potable water, garbage disposal and adequate chilled storage space.

Requirements for food business premises are specified in the Code (Standard 3.2.3 Food Premises and Equipment). Additional guidance is provided in Safe Food Australia and the Australian Standard, Design, Construction and Fit-out of Food Premises (see page 271).

Local council and/or state or territory authorities are responsible for ensuring food businesses meet the requirements specified in the Code. In some states or territories, before granting approval to commence selling food products, an Environmental Health Officer (EHO) may come to inspect the premises.

Requirements commonly found lacking during EHO inspections of small businesses are:

- Bench space is not large enough to safely prepare food – working in cramped and crowded conditions leads to multiple food safety hazards, including an increased risk of cross contamination.
- Refrigerated storage space is not sufficient – overcrowding of fridges reduces their ability to chill foods quickly or to maintain an adequate temperature, and prevents proper separation of raw and cooked foods.
- Temperature or other measuring equipment is not available – this leads to 'guesswork' regarding key food safety controls, such as adequate heating and cooling.
- Hand washing facilities are not provided – food handlers can transfer pathogens from themselves to food (e.g. not washing hands after using the toilet) or from one food to another (e.g. from raw meat to cooked meat).
- Control over flows of food are lacking – raw food should always be separated from ready-to-eat food. If there isn't a system in place to help achieve this (e.g. designated food preparation areas), there is a higher risk of pathogens from raw food contaminating ready-to-eat food.

Home-based food businesses

Private residences can be used to prepare food products, but you have to be able to demonstrate to an EHO that you can do this without jeopardising the safety of the products. The EHO will inspect your kitchen and other facilities and decide if they are appropriate for the type of products you plan to make. You may be required to make some alterations (e.g. increasing your bench space) or purchase new equipment (e.g. a larger fridge). In some cases, you may not be permitted to prepare certain high-risk foods in a domestic kitchen. The EHO will also be able to tell you about any licensing or registration requirements specific to your state or territory food authority.

Although home-based kitchens must comply with all relevant parts of the Code, the EHO will take into consideration the type of products you intend to make and the food safety hazards associated with their production. For example, the EHO may be more likely to approve the production of fudge in a domestic kitchen than if you plan to make chilled baby food.

As well as a physical inspection, the EHO will ask you questions relating to how the residence will be shared with others in your household. For example, how will you exclude young children and pets from the kitchen when food is being prepared, or what will you do if someone in the household shows symptoms of a foodborne illness?

The EHO may grant those operating a food business in a private residence exemption from meeting specific requirements of the Code. One example is exemption from the requirement to have hand washing facilities located within food preparation areas. This may be granted if there is a suitable hand washing basin that is easily accessible from the kitchen.

Suitability of existing business premises

You may already be operating your food business in an existing premise. This may have been built several years ago, and is now suffering from wear and tear, or it may not have originally been designed as a food business. Alternatively, it may be a multi-purpose food business such as a retail store or a food service outlet. Examples include delicatessens, fruit and vegetable shops, cafés and restaurants. If an EHO is satisfied you can prepare your products safely, you may not be required to make any major alterations to the premises. On the other hand, if you choose to make any alterations or structural changes to the building, it is advisable to speak to an EHO very early in the planning stages, in case your compliance is affected.

New food business premises

Designing your own premises gives you the opportunity to really think about the ideal environment to prepare your products in, both from a food safety and 'user-friendly' point of view. Guidance can be found in the previously mentioned Australian Standard, Design, Construction and Fit-out of Food Premises. Advice should also be sought from an EHO at the early planning stages, because there may be several different sets of local, state or territory regulations that must be met before being granted approval for building to commence.

Layout of premises

To reduce the risk of food becoming contaminated, you should carefully consider where you perform certain activities or locate particular facilities. Some activities and facilities should be physically separated from others, for example:

- Raw-material handling areas, including where deliveries are received, should be separated from processed product handling areas. It is important to separate raw meat handling areas from areas where ready-to-eat foods and food contact packaging are handled.
- Dishwashing areas should be separated from food handling areas.
- Staff amenities, such as toilets or change rooms, should be separated from food handling areas.

The ideal floor plan for a food business positions the raw material receipt, processing, packaging and storage areas so that food only moves, or flows, in one direction. However, this may not be achievable unless the premises have been purpose-built. To control the direction of food movement in existing premises, internal walls or partitions can be added. The illustration on the following page shows how adequate physical separation can be achieved (the red arrows indicate the direction food flows).

Other factors linking food safety and floor plans are:

- Doors to the outside of the building, or to toilet facilities, should not open directly onto food preparation areas, particularly those where cooked or ready-to-eat products are left uncovered for any period of time. This reduces the risk from airborne contaminants.
- Pipe work for sewage or waste water should never be located above food preparation or food storage areas. This reduces the risk of food becoming contaminated if leaks occur.
- Pipes carrying clean water should preferably not be located above these areas. This means that drips from condensation formed on pipes cannot contaminate food with pathogens potentially living in the moist environment on the outside of the pipe.
- Essential facilities should be easy to access; if they aren't, the chances of staff forming bad habits are increased, for example:
  - if cold rooms are difficult to get to, staff may be tempted to wait until there are several items needing to be returned to the cold room, instead of taking them straightaway
  - if hand washing facilities are not conveniently located close to food preparation areas and toilets, staff may feel they have an excuse for not washing their hands.

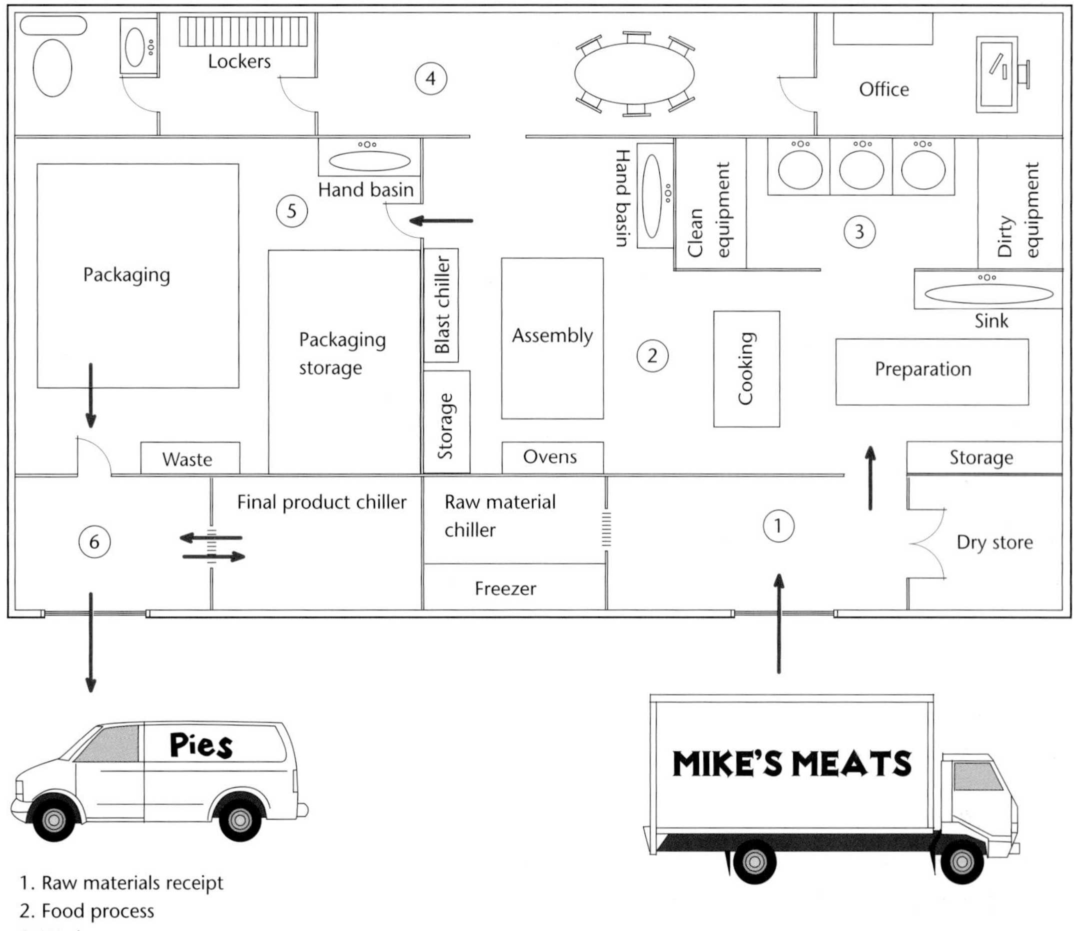

Physical separation of areas and direction food flows

The size of premises

It is not possible to specify requirements for the minimum size of a food business premises because each business is unique. Businesses prepare different types of products in varying quantities and have differing needs for equipment and staff numbers. Before choosing your premises or deciding to start your business from home, you should be able to estimate your space requirements. The Code requires that there is adequate space for food handling activities, including space for the equipment and other items required to do those activities.

Some of the things you will need to 'fit in' to your premises include:

- space to allow separation of raw and cooked or ready-to-eat foods during handling and storage
- separate refrigerated, frozen and dry goods storage areas for ingredients and final products awaiting distribution
- storage space for recalled products or items on hold
- separate sinks for washing up dirty utensils and equipment, and for washing and/or sanitising vegetables (if required)
- storage for clean and dirty uniforms, staff changing rooms and staff storage lockers
- room for equipment, such as industrial-size ovens and stoves, freezers and fridges
- storage for cleaning equipment and chemicals
- staff toilets and hand washing facilities, in addition to hand washing facilities in the food preparation area
- additional or larger volume hot water supply unit
- product packaging storage area
- garbage and recyclables holding area
- office space for paperwork, documents and records.

Garbage and recyclable material holding area

Food businesses may generate large volumes of food and packaging waste. Waste must be adequately contained while it is on your premises before collection so that it does not become a food safety hazard (e.g. by attracting mice or rats).

Your local council or relevant authority will be able to provide you with information on collection of waste from commercial premises in your area; generally, collection will occur more frequently than from domestic premises. You can also arrange for appropriate waste containers (e.g. 'wheelie' bins) to be delivered to your premises if they are not available currently. You must have a suitable area outside your premises where these containers can be stored and this area needs to be kept clean. The waste-storage containers must also be designed for easy and effective cleaning.

If you live in an area where waste containers are not supplied, you must provide your own. Those used for food waste must be leak proof and have sealable lids so pests are not attracted to your premises.

Air circulation or ventilation systems

Any food preparation, processing or cleaning activities that generate steam can cause condensation to build-up on ceilings, walls or equipment. Pathogens can live and grow in this moist environment. Condensate can drip off the ceiling or run down walls, potentially contaminating uncovered food or food contact surfaces with pathogenic microorganisms. Extraction and ventilation systems should be used to reduce or prevent condensation. Some pieces of equipment, such as commercial dishwashers, may have their own built-in extraction system requiring venting to the outside. You should contact a licensed tradesperson if you need to install extraction fans or exhaust hoods in your premises.

Having adequate air circulation in areas where steam is generated will also help prevent condensation forming. This can be via natural (such as open screened windows or doors) or mechanical systems (such as ducted ventilation). Ducted ventilation systems use a combination of air inlets and outlets fitted into the ceiling space, which are controlled by fans.

Too much air circulation can also be a food safety issue. Businesses that handle high-risk products, such as ready-to-eat foods, must pay particular attention to this issue. Doors or windows that open directly to the outside should be kept shut in areas where these products are handled. This reduces the likelihood of pathogens or insects entering the premises via incoming air. It also provides greater control of the direction and strength of airflows within food handling areas, which are needed to reduce the risk of cross contamination from raw to cooked food. After all, there is no control over the strength of air coming in via an open door or window.

In premises where both raw and cooked foods are handled in the same room, the direction that air flows within the room can be controlled by correct placement and operation of air vents. You should know what the pattern of airflow is between inlet and outlet vents in the ceiling. Ready-to-eat foods should be handled upwind from where raw foods are handled (i.e. air flows from ready-to-eat to raw only). Under no circumstances should fixed overhead or portable fans be used in areas where both raw and ready-to-eat foods are handled at the same time.

If your premises have separate rooms for handling raw and ready-to-eat foods, you should try to have positive air pressure in the clean room where ready-to-eat foods are handled. This means air is forced from the clean area to the less clean area when the door between them is opened. An air conditioning technician may be able to assist you to achieve these conditions.

If your premises has no built-in systems, and you leave doors open for ventilation, you must install adequate protection from contamination (see under 'Windows and doors' below).

Lighting

Adequate lighting plays a role in minimising food safety risks in food businesses. For example, if lighting is too dim:

- physical contamination of food may not be noticed

- food particles may not be adequately removed from surfaces and equipment during cleaning
- temperature gauges may not be seen clearly
- signs of pest infestation may be missed.

If an EHO observes lighting levels are inadequate in existing premises, you may be required to rectify this by upgrading. For example, you can add in extra light fixtures or increase the wattage of globes in current fixtures.

If light fixtures, globes or tubes located in food preparation or storage areas contain glass, the risk of broken glass contaminating food should be minimised by:

- shielding with plastic covers or wire meshes
- fitting with non-shatter globes.

Floors, walls and ceilings

The key requirement for floors, walls and ceilings in food business premises is that they are made of smooth and durable material that is able to be effectively cleaned and, if necessary, sanitised.

This means they should be:

- free of damage such as cracked tiles, holes in grouting or flaking paint
- able to withstand the application of cleaning agents such as hot and/or high-pressure water, steam or chemicals.

Tiles, vinyl sheeting or stainless steel are more durable than paintwork and are preferred walling options in areas that require frequent cleaning. These materials are also better for walls near sinks or equipment that creates steam because they are able to withstand higher temperatures and moisture levels.

Build-up of dirt and food residues on the floor will attract pests and harbour pathogenic microorganisms. Carpet pieces and other types of absorbent floor matting are not acceptable in food preparation, storage or washing up areas because they are not able to be effectively cleaned. Anti-fatigue and anti-slip mats used in food premises must be able to be cleaned effectively. Tiles or resin-based seamless flooring are appropriate flooring for food businesses.

If wet washing, such as hosing-down of floors, is performed, appropriately positioned drains and graded slopes are required to prevent pooling of water, which is a potential source of microbial contamination.

Windows and doors

External doors and windows should seal well when closed to reduce the risk of airborne contamination and pests entering the premises. Unless they are always closed, they must be fitted with adequate protection such as:

- fly-screens
- strip curtains (plastic flaps)
- air curtains – these blow air from top of door downwards, acting as a barrier to flying insects.

Screen doors should preferably be self-closing to minimise the length of time they are open. Large shutter doors that may be left opened, such as when unloading deliveries, can be fitted with either strip curtains or air curtains.

Fixtures and fittings

Fixtures and fittings include items such as benches, shelves, cupboards, sinks and light fittings; in general, these items are fixed in position.

Materials used for fixtures and fittings must be suitable for food preparation activities. Wood is not recommended as a food contact surface because it can absorb liquid, can splinter, and is

difficult to clean effectively and sanitise. Avoid the use of glass in food preparation or storage areas whenever possible, because of the chance of broken glass contaminating food. Stainless steel or plastic are recommended.

Benches used as food contact surfaces (i.e. unpackaged food is placed on them) must be capable of being effectively cleaned and sanitised. This requires bench tops to be smooth, and free from cracks and crevices. Stainless steel is recommended.

Shelves should be at least 150 mm off the floor to allow access for cleaning underneath. Although not ideal, shelving used for dry ingredient storage, can be made of wood if it is well sealed.

Racks used for draining equipment and utensils after washing should be made of stainless steel or plastic, and be large enough to accommodate all items.

If a **sink** is used for washing utensils and equipment, it should be at least double-bowled to allow separation of the cleaning and sanitising steps. Ideally, there should be three sinks, one for rinsing, one for washing and one for sanitising. Sinks used to clean and sanitise equipment should be big enough to allow for full immersion of large items such as chopping boards. If fruit and vegetables routinely need to be washed, a separate food preparation sink should be installed.

Hand washing facilities must be located where they can easily be accessed by food handlers. Facilities that are behind doors, obstructed by pieces of equipment or are positioned too high or low for the average person to use are considered to be inaccessible. If any food or items that will come into contact with food, such as packaging or equipment, are likely to be handled directly by staff, you must provide hand washing facilities within the areas where this work is performed. As a guide, a food handler should not have to walk more than 5 metres to the nearest hand washing basin. Hand washing basins should preferably be supplied with liquid soap and paper towels. In general, air-driers take too long to dry hands and discourage staff from drying their hands completely.

Hand washing facilities must be provided immediately adjacent to the toilets. These facilities are additional to those located in food handling areas.

Sinks used for other purposes, such as washing vegetables, should not be used as a hand washing sink, and vice versa (except if an exemption is granted, see below). Your hand washing sink should be easily recognisable as being solely for washing hands. This may be achieved by installing a sink that looks like a typical hand basin and putting up a sign that says 'Hand washing only'. A sign that shows a picture of hands being washed will add emphasis and act as a reminder to staff (see the photo below).

Small businesses with limited resources, or those operating food premises that have limited space, may be permitted to use one side of a double-bowled sink for hand washing. The EHO will take into consideration the types of activities performed at the premises before making their decision.

Hand washing facilities must supply warm, running potable water. Most Australian town or mains water supplies are potable, but it is your responsibility to confirm this with your local council or relevant authority.

Hands-free taps, which are turned on and off via motion sensors, by elbow-operated levers or by a foot-operated pedal, can be installed. However, there is no requirement in the Code to do so, and the same level of hygiene can be achieved using conventional taps.

Storage facilities must be provided to segregate any non-food items, which may be a source of food contamination, from food. Storage in a room separate from food preparation or food storage areas is preferable, but, if this is not possible, lockers or cupboards should be designated for these items and, if necessary, locked.

Items include:

- chemical storage cupboards – for bulk chemicals used for cleaning, sanitising, pest-control and equipment or vehicle maintenance
- staff lockers or cupboards – for personal belongings such as clothing, handbags, jewellery and food
- cleaning equipment room or cupboard – for items such as brooms or mops and diluted cleaning and sanitation chemicals
- dirty linen basket – for soiled uniforms, aprons or cleaning cloths
- storage for miscellaneous potentially hazardous items – such as stationery (e.g. paperclips and staples) and building maintenance items (e.g. nails and paint).

Toilet facilities must be available for staff use. Toilets may be located within the food business premises or they may be available as part of the complex the premises are in (e.g. a shopping centre). Regardless of their location, they must be working properly and include hand washing facilities supplied with warm, potable running water, soap and suitable hand-drying facilities.

Doors to toilet cubicles should not open directly onto food handling areas. There must be a vestibule or similar space to act as an airlock between the two environments.

Thinking ahead can prevent future problems

To stop building and equipment maintenance issues becoming food safety issues, a preventative maintenance program is necessary. Preventative maintenance means action you can take in advance to stop occurrence of a problem in the future.

For example:

- changing a filter in an exhaust fan regularly, rather than waiting for it to become so dirty that it no longer works effectively and condensation starts to build-up on the ceiling
- regularly inspecting drains so any minor partial blockages can be cleared before drains become completely blocked and overflow

- servicing equipment regularly so that bolts that have worked loose can be tightened to stop them falling into food or causing equipment to break down.

Any repairs required in your premises should be performed with minimal delays. Putting off repairs will generally only make the defect worse and can put your product at risk, taking more time and money to fix. For example, one tile lifting on the floor may allow water to seep underneath during cleaning, leading to further loosening of tiles and creating a moist environment for pathogenic bacteria to grow.

Fit-for-purpose equipment

The equipment you use in your food business must be of a suitable design and in good condition to reduce food safety risks.

Purchasing equipment

One of the key food safety considerations when purchasing food processing equipment is how well it can be cleaned and sanitised, and how easy this process will be. Smooth surfaces (e.g. stainless steel) are better than rough (e.g. plastic that has a pattern); wood should be avoided for all food contact surfaces and utensils. Surfaces should be resistant to cleaning and sanitising chemicals used at their recommended concentration.

It is a good idea to inspect equipment thoroughly to see if there are any crevices or hard-to-reach spots where food and grime can accumulate. You can ask the salesperson to demonstrate the process of dismantling pieces of equipment for cleaning and sanitising, such as pulling off a cover or guard. Some equipment comes with specialised cleaning tools, which you should be shown how to use correctly.

Also consider how easy it will be to dry the equipment after cleaning and sanitising. Because *Listeria monocytogenes* and other pathogens like to live in moist areas, drainage holes for releasing water from inside equipment should work effectively to avoid pooling.

Equipment should be suitable for its intended use, for example:

- if you handle any potentially hazardous food, you must have a thermometer accurate to within 1°C because these foods have to be stored under strict temperature control (Boxes 22 and 23)
- dishwashers used to sanitise equipment should have a cycle with an adequate time–temperature combination to sanitise effectively (see Box 31 under 'Cleaning and sanitising' below)

- fridges, freezers and/or cold rooms should be large enough to meet your businesses requirements
- equipment such as cookers or blast chillers must be able to heat or cool food quickly enough to prevent microbial growth (see Chapter 6 for recommended times)
- garbage bins used in food preparation areas should ideally have 'hands-free' lids, such as a foot-pedal operation.

If you cannot find equipment that will allow you to process your food safely, you will have to change how you make your products – you cannot simply 'make do'. For example, if a blast chiller capable of cooling your intended product batch sizes within the required time is not available you must decrease your batch size or use an alternate cooling method such as ice water.

Equipment installation and storage

Large, non-portable pieces of equipment should be installed so that you can easily clean around them (on the sides and underneath). You may also need to allow for the removal of covers that swing off in only one direction. Small gaps between walls and fixtures should be filled-in to stop food dropping down and building up. This attracts pests and/or provides a source of nutrition for pathogen growth.

It is best to place fridges and freezers away from equipment such as ovens that generate a lot of heat. Although they are insulated, this may still hinder their ability to cool effectively. You should also follow the manufacturer's guidelines for how much clearance there should be between the fridge or freezer and walls or other pieces of equipment, because poor airflow may also reduce their cooling capability.

Small items such as thermometers should be stored in a designated place. Instruct staff to return items promptly after use to reduce the risk of essential measurements not being taken owing to misplaced equipment.

Box 22 – What are potentially hazardous foods?

A potentially hazardous food is any food:

- that may contain pathogenic microorganisms, either naturally occurring or through contamination during handling, and
- is able to support growth of, or toxin production by, these pathogens.

In general, these foods are perishable with a low-acid content; examples include:

- cooked meat (including poultry and game) or foods containing cooked meat such as casseroles, curries and lasagne
- smallgoods, such as Strasbourg, ham, corned beef and chicken loaf
- dairy products, such as milk, soft cheeses, custard and dairy-based desserts
- seafood, including seafood salad, patties, fish balls, stews containing seafood and fish stock
- prepared fruits and vegetables, such as salads and pre-cut melons
- cooked rice and pasta
- foods containing eggs or beans, and other protein-rich foods such as quiche, fresh pasta and soybean products
- combinations of these foods, such as sandwiches, rolls and pizza.

Some ingredients that are initially not potentially hazardous may become potentially hazardous because of changes made by you. For example, dried powders such as milk or custard powder are too dry to support the growth of pathogenic microorganisms, but once they are added to water they must be treated as potentially hazardous.

If you are unsure if a food is potentially hazardous or not, you should check with an EHO.

Box 23 – Potentially hazardous foods and temperature control

The Code (Standard 3.2.2) defines a potentially hazardous food as: '… food that has to be kept at certain temperatures to reduce the growth of any pathogenic microorganisms that may be present in the food or to prevent the formation of toxins in the food.'

To keep these foods safe, they must be stored at a temperature that will prevent, or significantly slow, pathogen growth or toxin production:

- 5°C or below for storage of chilled foods
- 60°C or above for those foods intended to be served hot
- at another temperature for a limited time, if it can be shown this alternate combination is safe.

Keeping things humming along

If equipment breaks down halfway through a production run, the safety of your products could be at risk. For example, a partially processed product may be left for long periods at room temperature while repairs are made, potentially allowing pathogens to grow to unacceptable levels. Like your car, food processing equipment needs regular servicing. A preventative maintenance program for your premises and equipment should be in place to prevent break downs. This includes maintenance your staff can do and maintenance that is performed as part of servicing by professional contractors. A maintenance log should be kept so that you can show evidence that these tasks are being completed.

It is not necessary for equipment to stop working to become a food safety hazard; it may still be operating but not be achieving the correct parameters. For example, a heat sealer with misaligned jaws or vacuum packaging equipment may not be removing enough air.

Most equipment instruction manuals will include details of preventative maintenance required. The troubleshooting section may provide information about any signs your equipment may give to show that it requires attention.

Examples of different types of preventative maintenance for equipment include:

- tightening or re-positioning conveyor belts
- lubricating moving parts
- tightening screws and nuts

- replacing O-rings or gaskets
- replacing any parts that show signs of rust or corrosion
- replacing batteries in digital thermometers.

To keep on top of what needs to be done and when you need to do it, you should record details of your preventative maintenance program in a maintenance schedule. See Box 24 for an example.

You must make sure that staff or contractors that maintain and repair your premises and equipment know how to do their duties without creating any food safety issues.

This includes:

- use of 'food grade' lubricants – lubricants can become contaminated with pathogens; they should not be left sitting around with the lid off and they should be dispensed carefully
- keeping track of any small parts removed or tools used, so they do not accidentally make their way into your products
- not using food packaging to store screws, washers, tools or other non-food items
- following correct hygiene procedures, including hand washing and any specific instructions received upon arrival at the premises (e.g. areas where entry is not permitted).

After any maintenance activity, the equipment and surrounding areas should be thoroughly cleaned and/or sanitised.

In addition to a preventative maintenance program, you should encourage your staff to report anything unusual they observe when using equipment. Even something that may appear minor may indicate a food safety risk. For example, slight resistance when opening a cover may be caused by rusting hinge screws, which, if not replaced, may eventually break off and fall into food.

Equipment monitoring and calibration

If you simply assume your equipment is working properly without actually confirming this, you run the risk of undetected equipment faults compromising the safety of your products. For example, you may need to store potentially hazardous food at 5°C or below to control microbial growth. You have set the temperature of the cold room to 4°C, but the thermostat is faulty and the actual temperature inside the cold room is 7°C. If the thermostat is not calibrated annually by a refrigeration mechanic, you may not detect this discrepancy.

Box 24 – Example preventative maintenance schedule (equipment and premises)

SMITH & SONS

2010 preventative maintenance schedule

Month due	Week due	Equipment/ area	Task	Staff member/ company responsible	Date completed
Jan	2nd	Freezer room	Defrost and inspect seals	Brian	9/1/10
	2nd	Heat sealer	Inspect jaw alignment and adequacy of seal	Mary	8/1/10
	3rd	Strip curtains	Check for loose, damaged or missing strips	Brian	16/1/10
Feb	1st	Vacuum packager	Yearly maintenance	V Pack Pty Ltd	13/2/10 Technician could not make it last week
	2nd	Freezer room	Defrost and inspect seals	Brian	6/2/10
Mar	1st	Delivery truck	6 month service	Top's Auto Service and Repair	7/2/10

For food safety reasons, any item of equipment used for:

- heating food (e.g. ovens and retorts)
- cooling food (e.g. fridges or cold rooms)

must be monitored.

Other equipment or processes that may need to be routinely monitored to reduce food safety risks include the water temperature reached during the sanitising cycle of a dishwasher, the weighing accuracy of a set of scales, or the readings given by a pH meter (used to measure how much acid is in food).

There is little point in monitoring equipment if you do not know what an acceptable level of accuracy is. For example, when checking your scales you place a 200 g weight on them and read the value given. Would it be acceptable if the readout said 205 g or 210 g or 230 g? Having a range of values defined as acceptable, removes the need for individuals to make up their own mind. This range of values is called an acceptance criteria. In this example, the acceptance criterion for a 200 g weight may be 195 to 205 g (i.e. plus or minus 5 g).

If supplied, you should use information in equipment manuals or specifications as a guide to setting your acceptance criterion. Alternatively, you can seek advice from the manufacturer or company that supplied the equipment.

The Code (Standard 3.2.2) specifies the level of accuracy required for thermometers used to measure the temperature of potentially hazardous food be within 1°C (plus or minus). For example, if you were measuring the temperature of boiling water, which is normally 100°C, it would be acceptable if the thermometer readout was somewhere between 99 and 101°C.

You also need to be confident that the instrument you are using to monitor equipment is accurate itself. Calibration determines the accuracy of measuring instruments against a 'known' standard. Types of instruments that require calibration are thermometers, weights used to check scales and pH meters. Further detail on thermometer calibration and monitoring the temperature of cold rooms or fridges is provided in Boxes 25 and 26. Instruction manuals provided with pH meters will also have information on how and when they require calibration.

Like a preventative maintenance schedule, you should make a list of all the equipment on your premises that requires monitoring or calibration. It is also worth having step-by-step written instructions on how to perform this, including what the acceptance criteria are. Keeping records of the values collected in your checks is important so they can be referred to if there are any suspected food safety issues with your products. If you have more than one unit of the same piece of equipment (e.g. probe thermometers), you should label each with an individual code so they can easily be told apart.

Any equipment that fails to meet your acceptance criteria during routine monitoring or calibration should be serviced, repaired or replaced without delay. Some equipment can be adjusted to improve its accuracy, but, unless the manual describes how to do this, only a qualified technician should do the adjustment.

If you discover that a cold room, fridge or freezer is not working properly, or has suffered a complete breakdown, you have to act quickly. If the food inside the fridge or freezer is still at 5°C or below you can transfer it to another unit. If you do not have enough spare space in a fridge,

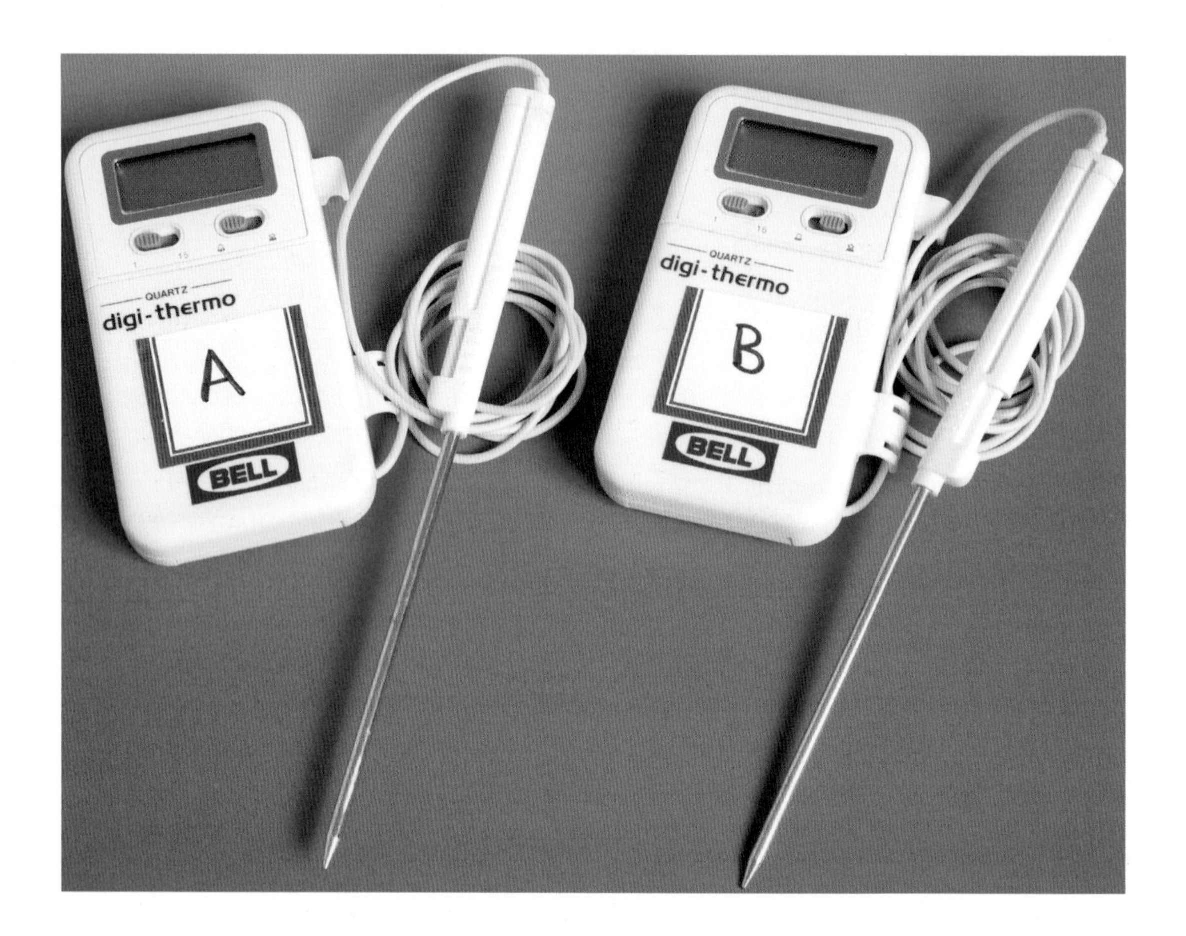

you may choose to freeze some foods. Refreezing thawed frozen food is not a food safety issue as long as the temperature of potentially hazardous food is kept at 5°C or below; however, the quality of some foods may be affected by this process. As a general rule, if any potentially hazardous food has reached 10°C or above, it is advisable to discard it. However, if you can prove that the food has been at the higher temperature for less than 2 hours then it is possible it may still be usable; refer to Chapter 5, Box 56 for information on how to determine this.

Box 25 – Thermometer calibration

Thermometers used to measure the temperature of:

- potentially hazardous foods and/or
- cold rooms and fridges potentially hazardous food is stored in

require routine calibration so they remain sufficiently accurate to meet the requirements of the Code.

Thermometers usually require calibration once a year. However, you should check with the thermometer supplier or in the instruction manual for the recommended frequency. Any thermometers appearing to give incorrect readings, or those that have been dropped or damaged, should be calibrated without delay.

A reference thermometer can be used for the calibration procedure. This is either a new thermometer calibrated by the manufacturer or a special thermometer calibrated by a laboratory or the supplier. The reference thermometer must only be used for calibrations, to stop it from being damaged and losing accuracy.

Most thermometers will need calibration at the hot and cold end of the temperature scale. The steps to follow are:

1. Place the thermometer to be calibrated and the reference thermometer in an insulated flask or thermos filled with very hot or boiling water (around 100°C). Immerse to the minimum depth specified in the thermometers instruction manual.
2. Gently stir the water with the thermometers and record the temperature reading for both.
3. Replace the hot water in the flask with a mixture of crushed ice and just enough water to fill all the air gaps. This should give a temperature reading around 0°C.
4. Gently stir the water with the thermometers and record the temperature reading for both.
5. Review the results.

Possible results and actions you should take:

- If the temperature readings given by both thermometers are each the same in the hot and cold tests, the thermometer being calibrated is operating accurately and no further action is needed.

- If the reading on the thermometer being calibrated is more than 1°C different (either plus or minus) from the reference thermometer for both the hot and cold tests, and this difference is the same for both tests, record the temperatures and mark the thermometer being calibrated with a correction factor.
- If the thermometer being calibrated is out by 1°C or more for the hot tests but reads correctly for the cold test (or vice-versa), you must have the thermometer being calibrated repaired or replace it with a new one. It is unsafe to apply correction factors in such cases.

Example A:

Test	Reference thermometer	Thermometer being calibrated	Correction factor
Hot test	100°C	98°C	plus (+) 2°C
Cold test	0°C	–2°C	

Example B:

Test	Reference thermometer	Thermometer being calibrated	Correction factor
Hot test	100°C	103°C	minus (–)3°C
Cold test	0°C	3°C	

Then every time the thermometer is used the appropriate correction factor needs to be added. For Example A above, if the thermometer reads 68°C when put into a pot of stew, you would apply the correction of plus 2°C, and know that the real temperature of the stew is 70°C.

You can only apply correction factors if the thermometer is out by the same amount for both the hot and cold tests.

If you do not feel confident enough to calibrate your thermometer yourself, you can send it out to a laboratory. It is preferable to have a second thermometer (also calibrated, but at an alternate time), so you can still continue to perform temperature measurements.

Box 26 – Monitoring cold room and fridge temperatures

If you store potentially hazardous food in any of your fridges or cold rooms, they must be able to maintain food at 5°C or below. You must monitor the temperature inside this equipment every day so you know you are achieving this.

All fridges and cold rooms are fitted with a thermostat, which turns the cooling fans or motor on and off to maintain the set temperature. Commercial units also have a link from the thermostat to an external digital display, showing you the temperature reading inside the unit. This reading should be used to monitor the temperature of these fridges or cold rooms.

If your fridge or cold room is not fitted with a temperature display, you will need to place your own thermometer inside. First, find where the warmest part of the fridge is, because this represents the 'worst case scenario'. To do this, place the thermometer in different spots each day (e.g. top, middle or bottom shelf), recording the temperature at each spot before moving it to the next position. Once you have found the area where the temperature reading is highest, leave the thermometer in that spot to use for your daily temperature reading.

Keeping a record of the daily temperature allows you to observe if there is a gradual increase over time (see Box 27 for an example). This may indicate that the unit needs defrosting, servicing, or is being over-filled.

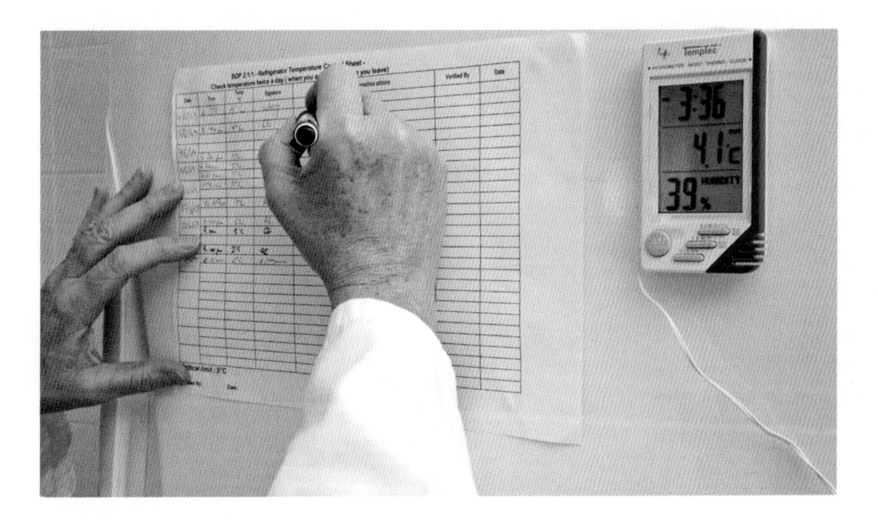

Box 27 – Example routine temperature monitoring

SMITH & SONS

DAILY TEMPERATURE LOG – COLD ROOM NO. 1

Method: Check the temperature before opening the door for the first time in the morning.

Acceptance criteria: The display should read 5°C or below. If the temperature is above 5°C, contact supervisor for corrective action instructions.

Month	Date	Reading (°C)	Acceptable	Initial	Corrective action
Apr 2009	1st	4	Y	BW	-
	2nd	4	Y	BW	-
	3rd	Business closed	-	BW	-
	4th	3	Y	BW	-
	5th	4	Y	BW	-
	6th	5	Y	BW	-
	7th	5	Y	BW	-
	8th	6	N	BW	Called supervisor who organised maintenance

Cleaning and sanitising

Effective cleaning and sanitation programs are required to achieve an adequate level of hygiene in food businesses.

A clean surface is free from visible food residues or particles, feels non-greasy to the touch and has no unpleasant odours. An effectively cleaned surface should also be virtually free from food allergens. A sanitised surface is a clean surface that is virtually free from pathogenic microorganisms.

Requirements of the Code

The Code (Standard 3.2.2) specifies that 'A food business must maintain food premises (including fixtures, fittings and equipment) to a standard of cleanliness where there is no accumulation of:

- garbage, except in garbage containers
- recycled matter, except in containers
- food waste
- dirt
- grease
- other visible matter.'

Food contact surfaces of equipment or utensils must also be sanitised (in addition to cleaning), '... whenever food that will come into contact with the surface is likely to be contaminated.'

The Code specifies that effective sanitising requires:

- application of heat or chemicals,
- heat and chemicals, or
- other processes

able to reduce the number of microorganisms on the surface or utensil to a level that does not compromise the safety of food and does not allow the transmission of infectious disease.

Routine cleaning and sanitising

Improperly cleaned or sanitised equipment increases the risk of food becoming contaminated by pathogenic microorganisms. This risk is even greater if biofilms build-up on your equipment or other food contact surfaces (see Box 28 to learn more). In addition, because routine cleaning prevents food scraps building up in your food preparation areas, insects and pests will not be attracted. If you make allergen-free products in addition to allergen-containing products, your routine cleaning and sanitising program will also need to reduce the risk of cross-contact contamination of allergens.

Floors, bench tops and used equipment require cleaning and sanitising at least once a day. However, some equipment or surfaces may need to go through this procedure more frequently. Examples of these include equipment that is known to have a high-risk for transferring pathogens (e.g. meat slicers), surfaces used for both raw and ready-to-eat food, or equipment that is used for both allergen-containing and allergen-free products.

Because sanitisers do not work well if there are food residues present, cleaning needs to occur before sanitising. However, if a product combining the actions of a detergent and a sanitiser is used, this 'rule' does not apply as long as the instructions provided with the product are followed and you are able to demonstrate effective sanitisation.

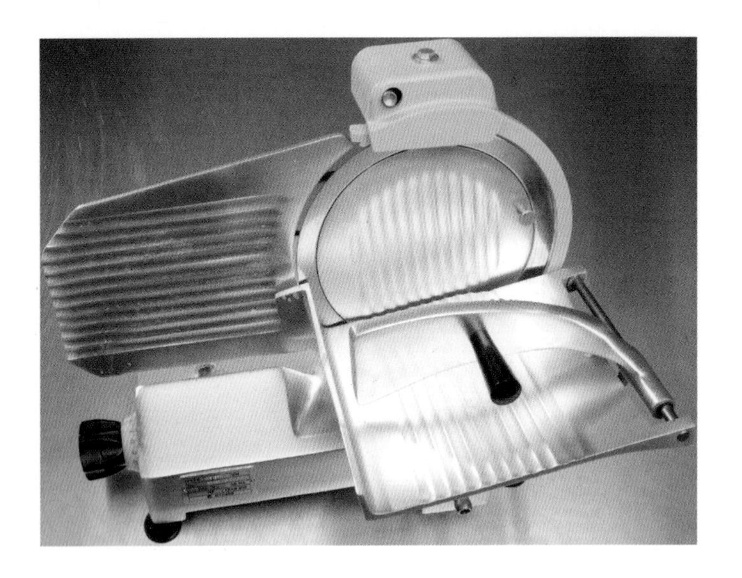

Box 28 – What is a biofilm?

Just as bacteria cause a build-up known as plaque on the surface of teeth, microorganisms can accumulate on surfaces in food preparation areas. These microorganisms form a type of community called a biofilm that protects them from cleaning and sanitising chemicals and hot water washing. Once formed, biofilms are very difficult to remove. Preferably, you should prevent them occurring in the first place by having a robust cleaning and sanitising program in place.

Here are the six basic steps for routine cleaning and sanitising:

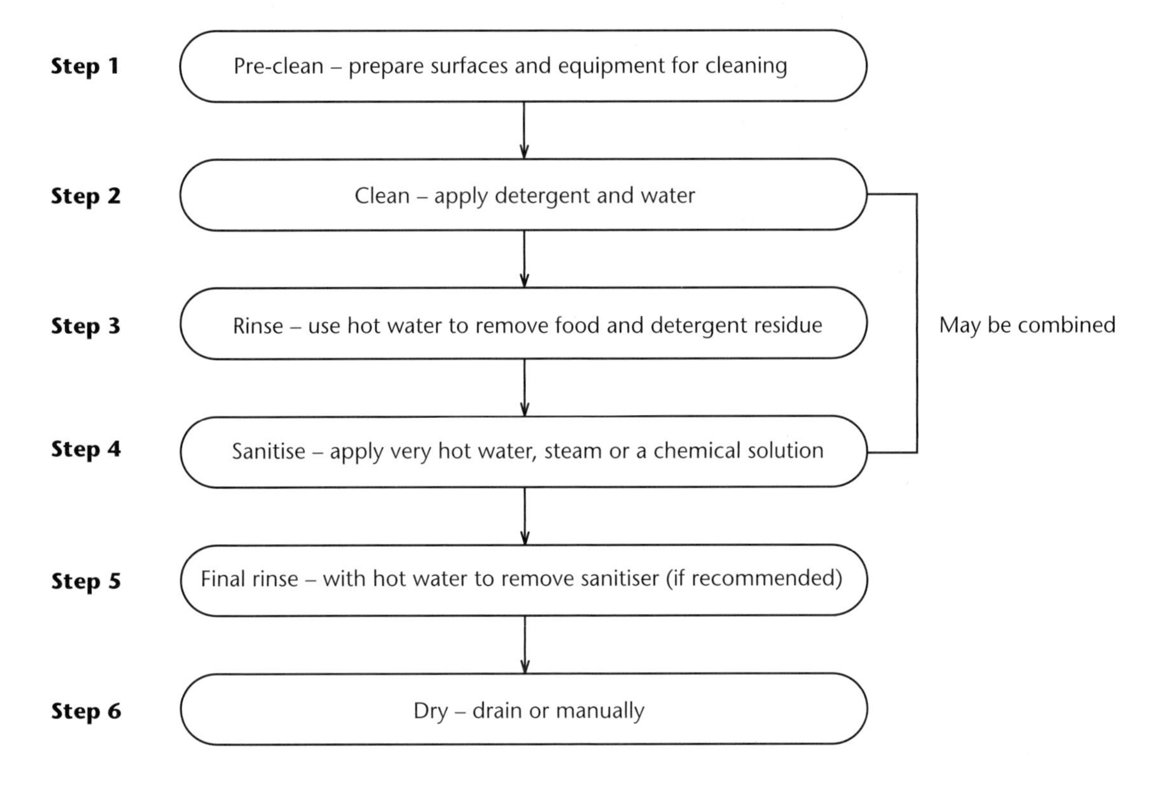

Step 1: Pre-clean – This prepares the surfaces or equipment for cleaning. Dry cleaning, such as brushing down equipment or sweeping floors, removes excess food particles. You may need to pre-soak or scrape items that have tough-to-remove food particles on them, such as 'baked-on' food.

Equipment that has 'nooks and crannies' where food particles may lodge, such as under O-rings, should be dismantled during this step: the manufacturers' instructions will provide guidance. Large pieces of equipment that can't be broken down into smaller parts should have covers or guards removed for access. Move portable equipment when cleaning floors, to remove drips and spills underneath them.

Step 2: Clean – Detergent, hot water and 'elbow grease' are used to remove food particles, fats and other residues (including allergens). Always follow the instructions provided with the detergent.

Bulky or fixed equipment will need to be washed down with detergent and hot water where it stands. Smaller pieces of equipment, utensils and other items can be immersed in a sink or tub. If the detergent loses its activity, or the water cools down too much, you will need to drain the sink and start afresh.

Step 3: Rinse – Rinsing is required to flush away food particles and detergent, providing a 'clean slate' for sanitiser application. This is an important process because if there are food residues present the effectiveness of most sanitisers is reduced.

Step 4: Sanitise – This is required to reduce the level of pathogens on food contact surfaces or equipment to safe levels. Equipment that comes into direct contact with food and floors in areas where high-risk foods are handled, such as ready-to-eat or cook chill, needs to be sanitised. Very hot water, steam or chemical sanitisers (also called disinfectants) can be used.

There are many types of chemical sanitisers available (see Box 29). These must be suitable for use on food contact surfaces and be used according to instructions supplied. Most will need to be diluted with water: too little sanitiser will be ineffective; too much may damage your equipment (e.g. metal corrosion). Chemical sanitisers require a minimum contact time, which is the time items need to be immersed in a sanitiser solution or the time left before rinsing off floors or benches.

The minimum temperature recommended for hot water sanitising is 77°C, with a minimum contact time of 30 seconds. However, if the water temperature is 82°C or above, the items do not need to be left soaking. These conditions are very difficult to achieve by immersing items in a sink (Box 30). Alternatively, a dishwasher with a sanitising cycle can be used for small items (Box 31).

Applying steam vapour directly onto surfaces is another sanitising method. Nozzles supplied with steam gurneys are useful for hard-to-reach spots. However, you must make sure there is no uncovered food around when you vaporise steam to prevent contamination.

Step 5: Final rinse – If the chemical sanitiser is not rinsed off items before they are used, it may affect the flavour of products. Some chemical sanitisers can damage equipment by accelerating corrosion, such as rust, if not rinsed off straight after use. However, some chemical sanitisers do not require rinsing off, or can be rinsed off just before you re-use the equipment, so always check the information provided with the product.

Step 6: Dry – Ideally, surfaces, equipment, utensils, and so on should be dried by draining and leaving exposed to the air. Racks should be available for draining items so they are not left sitting in pools of water. When possible, cleaned items such as bowls should be turned upside down to reduce the risk of physical contaminants falling into them.

Box 29 – Chemical sanitisers (including guidance for the use of bleach)

Each class of chemical sanitisers has their own advantages (e.g. non-corrosive or fast acting) or disadvantages (e.g. expensive or limited shelf-life once diluted). It is best to consult a chemical company to help you decide which products are suitable for your needs. It is your responsibility to check that the chemical you choose is safe to use in a food preparation setting (i.e. it is 'food grade').

A commonly used sanitiser is bleach, which contains an antimicrobial compound called chlorine (sodium hypochlorite). Those just starting their business may want to use bleach initially as it can be readily bought at any supermarket.

The chlorine concentration in bleach differs according to the type and brand. Bleach purchased from supermarkets is generally 4–5%; commercial or industrial strength bleach is usually 10%.

The amount of chlorine required to sanitise effectively depends on the temperature of the water used: the hotter the water the lower the concentration. However, it is simpler to use cold water for sanitising with bleach, because it is difficult to keep a sink full of water sufficiently hot.

Diluting bleach with water from the cold tap to make a mixture containing 0.01% (or 100 ppm) chlorine is an effective sanitiser for most situations. You should prepare a table showing how to make up this concentration for the specific product you have purchased. The table provided below shows you how to do this for two concentrations.

Items should be left soaking in the diluted bleach for 10 minutes. Because diluted bleach will lose its activity over time, diluted solutions should be prepared just before use. Undiluted bleach should never be used as a sanitiser: it is harmful to those using it and it will corrode equipment.

The amount of bleach required for different volumes of 0.01% chlorine

Household bleach (containing 4% chlorine)	Commercial bleach (containing 10% chlorine)
5 L water: add 12.5 mL bleach	5 L water: add 5 mL bleach
10 L water: add 25 mL bleach	10 L water: add 10 mL bleach
25 L water: add 62.5 mL bleach	25 L water: add 25 mL bleach

Box 30 – Sanitising with hot water

Unless a commercial dishwasher is used, it is almost impossible to achieve the requirements for effective sanitising with hot water (i.e. minimum 77°C, but 82°C or above is ideal):

- Because the maximum water temperature delivered by standard domestic hot water systems is approximately 40–60°C, a separate water heater would need to be installed.
- Placing items, especially large pieces of stainless steel equipment, into the sink will cause the water to cool faster.
- There is a risk of staff being burned by the hot water, particularly if splashes occur when placing items into the sink.

Box 31 – Using dishwashers for cleaning and sanitising

It is recommended that commercial dishwashers are used by food businesses; however, domestic dishwashers may also be acceptable.

Manufacturers and suppliers of commercial dishwashers sold in Australia must ensure their equipment is capable of thoroughly cleaning and sanitising. International standards used by manufacturers specify the requirements needed. This means when these dishwashers are used according to the instructions provided:

- items are thoroughly cleaned before the sanitising step
- the high water temperature applied for a specified time during the sanitising step effectively kills pathogenic microorganisms
- all items in the dishwasher are exposed to the cleaning and rinsing action of the dishwasher.

If you wish to use a domestic dishwasher to clean and sanitise items, you should seek advice from your local council or relevant authority. In some cases, the dishwasher may need to be connected to a water supply capable of delivering hotter water than your existing system. You should avoid using 'economy' or 'power-saving' modes on dishwashers because these generally have shorter cycles and/or lower hot water temperatures.

Dishwashers must be maintained in good condition so they operate effectively. This includes emptying filters and routine use of the cleaning cycle (or running a standard cycle while the dishwasher is empty).

Special indicators or temperature sensor labels that you pass through a dishwasher cycle are available. These change colour if the water temperature reached during the cycle is hot enough to sanitise. If you are using a dishwasher to both clean and sanitise, it is recommended that you use these to routinely check that your dishwasher is operating effectively.

If air drying is not always practical, paper towel or tea towels can be used. Tea towels must be clean and unused, otherwise they may be contaminated with pathogens (e.g. if they have been used to wipe dirty hands).

Periodic cleaning and sanitising

Even with a rigorous routine cleaning and sanitising program, food residues can build-up over time, particularly in hard-to-reach areas. Additional periodic procedures may be required, including:

- dismantling equipment more completely
- using a different detergent
- using a more abrasive brush or scourer
- using a chemical sanitiser instead of hot water or steam.

Circulation cleaning-in-place (CIP) can be used for routine cleaning and sanitising of equipment used to process liquid products (e.g. heat exchangers). Routine cleaning steps are performed using the equipment's automated flow system, at a high flow rate and pressure. However, the equipment will also need to be dismantled for periodic cleaning and sanitising because there can be hard-to-reach spots, such as beneath O-rings or in joints. Refer to the equipment manuals for recommended frequencies and procedures.

Monitoring cleaning and sanitising procedures

Once you have established your cleaning and sanitising program, you need to check it is effective. Staff should also let you know if they notice the methods they are following are not adequate and if changes are needed.

To check cleaning and sanitising is being performed adequately:

- do a visual inspection after cleaning to check for food particles or residues (perform this on a variety of items using a 'random spot check' approach)
- check the temperature of the water used for sanitising with a calibrated thermometer (in a sink) or temperature indicators (in a dishwasher)
- ask staff to demonstrate how they make up the correct concentration of sanitiser solution
- check if timers used to monitor sanitiser contact times are operating correctly
- review cleaning records on a regular basis to cross-check that all tasks are completed as required.

Confirming effectiveness of a cleaning and sanitising program

You should confirm if your program is consistently achieving its purpose. This process ranges from simple testing methods that can be done in-house, to more complex testing that may need outsourcing. Food businesses that prepare higher risk products will need to perform more frequent and rigorous testing, such as swabbing methods.

To confirm if cleaning is effective you should 'look, touch and smell'. For example, use a torch to look inside equipment that can't be fully dismantled, run your finger over stainless steel surfaces to feel for greasy residues, and check for unpleasant odours. Another method is to run swabs that detect protein over surfaces and if the colour of the swabs changes then food residues are still present. These checks should be done before sanitising to reduce the risk of re-contaminating items after they are sanitised.

To confirm if sanitising is effective you need to test for the presence of microorganisms. This can be achieved by running special swabs over sanitised surfaces and then testing the swab.

Two different types of testing are:

- Microbial swab testing – swabs are sent to a laboratory and tested for the total number of microorganisms present, or for specific types. Acceptance criteria should be established for levels and/or type of microorganisms detected.

- ATP swab testing – swabs are inserted into test meters, which check the level of adenosinetriphosphate (ATP) present. Presence of ATP indicates the presence of microorganisms or food residues. The equipment provides a pass or fail result for the swabbed area according to pre-programmed criteria.

If you are required to remove food allergens from equipment and food contact surfaces, you must also confirm that your cleaning and sanitising program is doing this effectively. There are food allergen test kits available that do not require any specialised equipment or training. Additionally, you may have your products tested for the presence of specific allergens. For example, if you use equipment to process a product containing peanuts, followed by cleaning and sanitising, then you use the same equipment to process a product not containing peanuts, you would send the first few samples from the non-peanut product run to a laboratory to test for any traces of peanut allergens.

If using swabs for testing, you should focus on testing food contact areas that are difficult to clean because they are most likely to trap contaminants.

It is recommended that you keep records of your test results and that you check these for trends over time. For example, although your results may consistently be within acceptable ranges, they may slowly be 'creeping up', indicating there is an issue with your cleaning and sanitising program that requires investigation and corrective action before it worsens.

Barriers to effective cleaning and sanitising

Effectiveness of cleaning and sanitation programs can be reduced by:

- inadequate staff training that can lead to use of incorrect methods or short cuts being taken because staff do not understand why cleaning and sanitising is important (e.g. not moving a piece of portable equipment to clean the floor underneath it)
- the production of tiny water droplets (called aerosols) during cleaning that can travel through the air and lead to cross contamination of an area that has already been sanitised
- blockage of floor drains, which can lead to water pooling and a growth environment for microorganisms
- poor equipment design that can cause water pooling or can trap food residues; problem areas include in-line temperature measuring devices, sharp elbows, dead spots or concave surfaces (see Box 32)
- overlooking less obvious areas that need cleaning and/or sanitising, such as equipment control buttons behind panels, light switches or door handles

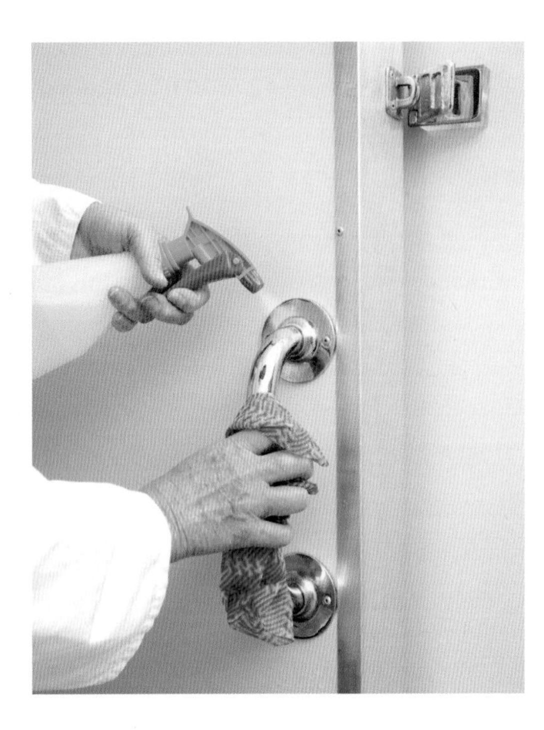

- incorrect use of sanitisers; for example, the contact time is too short, or the incorrect concentration or water temperature is used.

Timing and basic process

Ideally, cleaning and sanitising of larger pieces of equipment or surfaces such as walls and floors should occur when there is no food handling occurring. Cleaning can 'stir up' any foodborne pathogens and food allergens that may be present in the environment. If these become airborne, they can transfer to food or food contact surfaces.

Particular care should be taken with high-pressure hoses, dry sweeping and dusting. These cleaning methods (particularly use of high-pressure hoses) can spread aerosols over a great distance. These methods should only be used at the end of the day, followed by cleaning and sanitising all exposed food contact surfaces.

If a room or work area is to be deep-cleaned as a unit, this should occur in a logical sequence. Starting at the top and working down is the best method. For example, there is no point cleaning the bench tops before cleaning the ceiling.

Box 32 – Poor versus good equipment design

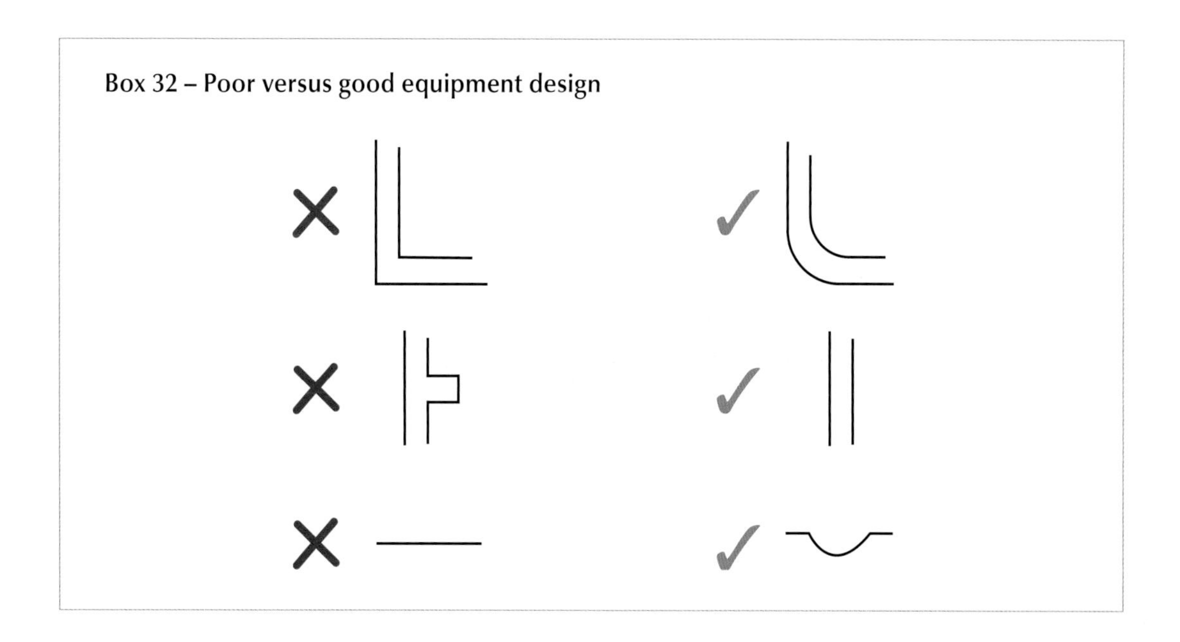

Most small businesses have few staff, so individuals may be required to 'multi-task'. In these circumstances, schedule cleaning tasks appropriately. For example, staff should only clean toilets after they have finished their food handling duties for the day.

Equipment used for cleaning

Equipment used for cleaning and sanitising should be kept in good condition so it does not become a source of contamination itself. For example, keep brushes and scourers clean and dry and replace as necessary to stop bristles or metal pieces contaminating food.

Cleaning cloths can be a food safety hazard. Wet cloths are a perfect breeding ground for pathogens. Disposable cloths or paper towel are best used for cleaning. Alternatively, have a plentiful supply of non-disposable cloths and instruct your staff exactly when and for what purposes they can or can't be used. Wash used cloths in boiling water and thoroughly dry at the end of every day.

If you use abrasive materials for cleaning, such as scourers, take care that they do not scratch your equipment or food preparation surfaces. Scratches make surfaces harder to clean and may trap food and microorganisms. It is preferable to use a different detergent or hotter water than to use harsh abrasives.

Cleaning and sanitising documentation

You need to keep track of what should be cleaned and sanitised, how often this should occur, who will do the task and how it will be done. Staff will also need to check-off when they have completed tasks so you can be certain nothing is being overlooked.

The first step in setting up documentation for your cleaning and sanitising program is to decide which items or surfaces need cleaning and which also require sanitising. The golden rule is that anything that comes into contact with food should be both cleaned and sanitised. However, there are some exceptions to this rule:

- items that will undergo a heat process only need to be cleaned (e.g. baking trays that are used in an oven)
- floors in high-risk areas (e.g. where ready-to-eat foods are packaged) should be cleaned and sanitised.

It is good practice to routinely walk through your premises to make sure nothing is overlooked, and then have members of your staff double check that you have covered off all items. It is easy to overlook something simple; for example, did you know that re-usable piping bags used for creams and custards (and other foods not cooked before eating) have caused several outbreaks of

Box 33 – Example cleaning and sanitising schedule

SMITH & SONS

CLEANING AND SANITISING SCHEDULE

Equipment/area	Frequency	Timing	Staff member responsible
Floors – processing	Daily	End of day	Brian
Floors – packaging	Daily	End of day	Brian
Bench top	Daily	As required and end of day	Mary
Meat slicer	Three times a day	8am, 11am, 2pm	Mary
Fridge 1	Weekly	Friday after stock rotation	Brian
Fridge 2	Weekly	Friday after stock rotation	Brian
Staff toilets	Daily	Last cleaning job of day	Mary

foodborne illness? They must be sanitised in addition to standard cleaning at the end of the day or, preferably, disposable piping bags should be used.

An example of documentation required to support your cleaning and sanitising program is provided in Box 33. This is for guidance only: the frequency and procedures required for your business will depend on the types of food handled and processing undertaken.

Pest control and animal exclusion

Pests, pets and farm animals can bring contaminants into your food business premises. Their presence must be controlled (pests) or completely excluded (pets and farm animals).

Pests are a food safety issue

Pests include birds, mice, rats, insects and spiders. The presence of pests in food business premises is a potential food safety issue because they can cause microbial or physical contamination of your ingredients and products. The Code (Standard 3.2.2) requires food businesses to take all practicable measures to stop pests entering the premises and to prevent pest infestations in the premises and in vehicles used for food transport.

Possible routes of contamination, either directly onto food or via transfer from equipment or food contact surfaces, are:

- bird, rodent or insect droppings, which can carry pathogens, or be a physical contaminant
- animal or insect 'feet', 'mouths' (e.g. when chewing on packaging) or other body parts, which can carry pathogens
- either whole pests or parts of them (e.g. insect bodies or bird feathers), which can cause physical contamination
- rats and mice, which will urinate continuously as they travel throughout your premises.

Pest detection

Signs that you have a pest problem at your premises include:

- rat or mouse droppings on the floor, benches or shelving
- signs of gnawing or teeth marks on food packaging
- cockroach droppings or tell-tale brown-black smears on surfaces
- evidence that baits have been eaten or disturbed
- urine tracks that show up under UV light.

Rats, mice or cockroaches are more likely to live in warm dark places, such as under fridges, near hot water systems or in dark cupboards. You may need to move equipment to check for the presence of these pests.

Keeping pests out and under control: in-house

The Code (Standard 3.2.3) states that 'The design and construction of food premises must, to the extent that is practicable, not permit the entry of pests and not provide harbourage for pests.'

Larger pests such as rats can be kept out of your premises primarily by having adequate physical barriers, such as keeping doors shut when they do not need to be open, minimising gaps under external doors or sealing off unused drains.

The preventative maintenance program for your equipment and premises should include:

- checking for holes or openings leading from the outside into the premises such as around pipes or ventilation grills; holes can be sealed up with cement, putty or commercially available 'gap fillers'
- inspecting weather strips used to reduce gaps under doors and replacing if necessary
- looking for holes or tears in screens on doors and windows, and repairing or replacing any that are damaged
- checking that self-closing units fitted to doors are working adequately.

Additionally, outside areas should be kept free of 'clutter' such as piled up pallets, cardboard for recycling or empty oil drums, as these provide nesting sites for rodents.

If there are bird nests or roosting spots outside your premises, especially if they are located in walkways, droppings on the ground beneath may be transferred into your premises on shoes. Ledges and other horizontal surfaces may need to be modified to discourage birds from 'taking up residence' in these areas. Birds are particularly hard to control and there are some ingenious traps and 'scarers' available on the market.

No matter how well you maintain your premises, the chance of these pests entering is increased if food waste held outside is poorly contained. Birds that scavenge food scraps, rats and mice are capable of breaking open plastic bags, so garbage bags must be put inside 'wheelie' bins or other containers with rigid lids. Any accidental spills of food outside bins should be cleaned up without delay.

Maintaining a clean waste area will also reduce the likelihood that pests such as cockroaches will be attracted to your premises. Similarly, an effective cleaning program inside your premises will make it less attractive for insects. It is particularly important to make sure any food residues or scraps on floors or other surfaces are cleaned up at the end of each day and that garbage bags are taken outside, even if they are not full, because many pests actively seek food during the night.

There are periods when doors need to be kept open. To deal with the flying insects that enter during these periods, you can install insectocuters or electronic insect 'zappers' with UV lamps to attract flying insects. These should be cleaned as part of your routine cleaning program (at least once a week) and placed where dead insects or parts of the unit cannot fall into food. The lamps lose effectiveness with time, so the globes should be replaced at least once a year. If you choose to use fly sprays instead, check the label for any warning statements about use in food preparation areas and make sure insects don't fall into the food when they die.

It is almost impossible to seal a building effectively enough to stop crawling insects from entering. It is also possible for insects to be brought into your premises on packaging or in food (including as larvae or eggs). The main aim is to make it harder for insects to breed and increase in number by stopping them from accessing sources of food and water.

So, in addition to the cleaning program mentioned above:

- ingredients should be stored in sealed containers
- food should be kept covered

- staff should be asked to take care with any food kept in their lockers or other personal storage areas (e.g. not to leave sandwich wrappings containing crusts or crumbs)
- dripping taps or other sources of water should be repaired.

Some pests, such as cockroaches, are best kept under control using regular preventative treatments rather than treating only after you see signs of their presence. This is because by the time you see evidence of cockroaches, they are probably already there in high numbers, making them harder to eliminate. Preventative treatments such as laying baits can be performed in-house or by a professional pest-control contractor.

Baits or traps should be laid at the first sign of rats or mice in your premises, even if you only suspect they may be present. Infestation with these vermin can become a serious food safety issue. If you believe that your in-house attempts to kill them are not working, you should call in a pest-control contractor immediately.

All pest-control chemicals must be stored separately from any food. Cockroach, mouse and rat baits must be positioned so there is no chance they will come into contact with food, equipment, food contact surfaces or packaging materials.

As with other programs required for food safety, it is advisable to prepare documentation for your in-house pest-control program to allow you to keep track of what needs to be done, how often, the method and who is responsible. A simplified example form is shown in Box 34.

Pest-control contractors

Pest-control contractors may be used on an irregular basis to fix infestations that you find you are not able to control effectively in-house or they can be used regularly to provide a preventative treatment program.

If using a contractor for routine preventative treatments, you should negotiate a contract that suits your needs. This will include:

- how often they will come – which is dependent on the treatment type
- what treatments they will apply – which pests they treat and how they do it
- a map of where they will spray or lay baits
- which areas they will inspect for evidence of pests
- what action they will take if a pest infestation is discovered.

Before using a pest-control contractor, you should check they have a current licence or registration. Additionally, contractors should supply written details, such as an MSDS, of any

Box 34 – Example pest-control program

SMITH & SONS

PEST-CONTROL SCHEDULE

Task	Frequency	Timing	Responsibility
Garbage area clean and tidy	Daily	Every afternoon	Brian
All foods covered in dry store	Daily	Every afternoon	Mary
Check and if necessary replace sticky cockroach baits	Fortnightly	Friday afternoon every 2 weeks	Mary
Check and clean electronic fly killers	Fortnightly	Friday afternoon every 2 weeks	Mary
Check external baits for activity	Monthly	First Monday of every month	Mary
Check for signs of pest activity in other areas	Monthly	First Monday of every month	Mary
Replace external baits (increase frequency if more activity observed)	Quarterly	First week of Feb, May, Aug, Nov	Mary
Replacement of electronic fly killer UV bulbs	Yearly	First week of Feb	Mary

poison used and prepare a report for you following each visit. These documents should be kept on file with your other records.

Pet and farm animal exclusion

There are only two scenarios in which live animals may be permitted in the premises of any food business (The Code, Standard 3.2.2):

- Live seafood, fish or shellfish – these are permitted in areas where food is handled. This allows food businesses to use fresh mussels, oysters and scallops, which are generally still alive when they are used.
- Assistance animals – this exemption relates to animals such as guide dogs for the visually impaired being allowed into dining and drinking areas. As such, this is not applicable to those producing manufactured or processed products.

Those operating their food business in their own home or at a farm (e.g. cheese making at a dairy farm) must have adequate barriers preventing domestic or farm animals from accessing any food handling areas. Because food transport is part of food handling, live animals are not permitted in food-transport vehicles. This should be noted by anyone just starting their business that may be using the 'family car' for picking up supplies and delivering products.

Food safety – the responsibility of all who operate or work for a food business

No matter how safe your premises are, your products are at risk unless food safety is a high priority for both you and your staff. Food safety should not be seen as an annoying 'extra thing to do' on top of routine activities; rather it should be an integral component of day-to-day activities.

Overall accountability for the safety of the products sold by a food business lies with the owner or owners of the business. Although staff have a responsibility to follow instructions and to perform their duties using hygienic practices, it is their employer's responsibility to provide the correct instructions, training, adequate equipment and facilities, and raw ingredients that are suitable for their intended use.

One basic, yet effective, way of showing staff that you take food safety seriously is leading by example. Your actions and attitude will guide the actions and attitude of your staff. Wash your hands regularly throughout the day, re-refrigerate leftover ingredients straight away, follow recipe instructions carefully, and so on. Your staff will then know what is expected of them.

Staff experience and training

The Code states that: 'A food business must ensure that persons undertaking or supervising food handling operations have – skills in food safety and food hygiene matters; and knowledge of food safety and food hygiene matters, commensurate with their work activities' (Standard 3.2.2).

This means that any of your staff who are classed as food handlers and those supervising them, have to:

- be capable of performing tasks needed to ensure the safety of the food they are handling, and
- know and understand why these tasks need to be done to make the food safe.

The skills and knowledge need to be relevant only to the tasks individuals perform (i.e. commensurate with their work activities). For example, a cook must know how to correctly measure the temperature of the soup they have prepared (a skill), and they should know this is required because a specified temperature needs to be reached to kill certain types of pathogenic microorganisms (knowledge).

Some Australian states or territories require that all licensed or registered food businesses have a Food Safety Supervisor (FSS). An FSS must be able to demonstrate to a Registered Training Organisation (RTO) that they have the specific competencies required for an FSS working in their sector of the food industry. This can be achieved by attending training run by an RTO and receiving a statement of attainment, or by showing that prior experience or training has allowed them to gain these competencies.

You must find out what your state or territory food authorities' specific requirements are regarding the skills and knowledge clause of the Code. Information provided in this chapter is only a guide to the requirements specified in the Code and the guidance provided in the FSANZ publication Food Safety: Guidance on skills and knowledge for food businesses (see page 272 for further information).

Making sure that your staff have a sound understanding of food safety and hygiene will not only keep your products safer, but can also be good for your business. Customers are becoming more food safety savvy and, if they observe your staff handling food in an unsafe way, they may be put off from purchasing products prepared on your premises. For example, if staff behind the counter of a delicatessen are seen touching food with their bare hands after wiping their hands on a dirty cloth, products that are advertised as 'homemade on the premises' may be unpopular.

Just as you will need to give other basic instructions to new staff, such as how to order supplies, food safety and hygiene skills and knowledge specific to their duties should be incorporated into all up-front training given to 'new-starters'.

This training does not have to be provided in a formal course. You may choose to use an 'on-the-job' approach such as using an experienced staff member to pass on skills and knowledge through verbal instructions and demonstrations. Encourage your staff to ask questions if they do not understand something they are told. Closely supervise any new staff members until they show they can perform their duties using the appropriate hygiene level and food safety controls. You should then also occasionally observe their food handling practices and check for bad habits, particularly during busy periods when they may be tempted to take short cuts, such as reducing cooking times.

Formal training may be a better approach in some circumstance; for example:

- a new business whose staff have no previous relevant skills and knowledge
- a business that is diversifying into a new area, such as a restaurant that plans to start a new side-line manufacturing products for sale in retail outlets
- a business manufacturing high-risk products, such as cook chill baby food or unpasteurised fruit juice.

It may only be necessary for one or two staff members to attend this training and they can then pass on the relevant skills and knowledge to others in your business. You can also hire a consultant to come to your premises and deliver training specifically tailored to your needs. Details of how to find appropriate training providers, including an RTO, are provided at the end of the book. Other ways your staff can learn skills and knowledge is through educational pamphlets, posters and DVDs.

Staff who attend a training course, or perhaps even watch a professionally made DVD, will also gain a better appreciation of why you ask them to do tasks in a certain way. That is, they learn the reasons why they need to use safe practices so they know it is not just based on your own personal preferences to have the job done 'your way'.

The health of food handlers

Pathogenic microorganisms can be transferred to food by food handlers suffering from a foodborne illness or condition. Common symptoms of foodborne illnesses are described in Box 35. Additionally, people can be carriers of a foodborne illness without having any symptoms of the illness or they can be carriers of a foodborne illness after they have stopped suffering symptoms of the illness. In all cases, the pathogens that are in their systems can still make others sick if they are transferred to food during handling.

If a member of your staff suspects that they have a foodborne illness or condition, are carriers of an infection or have come into contact with someone with a foodborne illness, they must inform their supervisor. A carrier of a foodborne illness is a person who is infected with a microorganism that can be spread to another person via food. However, the Code (Standard 3.2.2) states that a 'carrier of a foodborne disease does not include a person who is a carrier of *Staphylococcus aureus*'; to see why, read Box 36.

Box 35 – Common symptoms of foodborne illnesses

Diarrhoea and vomiting are the two most common symptoms of foodborne gastroenteritis. Staff who suffer from these symptoms and do not know what is causing them should suspect they have a foodborne illness. Other symptoms of foodborne illnesses are a fever, a sore throat with fever, or jaundice (yellowing of skin and whites of eyes).

If there is another known reason to be suffering any of these symptoms, staff are not required to report their symptoms to their supervisor. Examples of other causes of these symptoms are a bowel disorder or non-foodborne infection.

What occurs after a supervisor is informed will depend on the nature of the illness or condition and the type of work normally performed. Staff members may come into work but perform alternate non-food handling duties until they are well (e.g. filling in paperwork), or they may be required to stay at home until they feel well again. In both cases, they will be required to get advice from a doctor stating they are fit to recommence food handling before returning to normal duties. This is a requirement of the Code (Standard 3.2.2).

If staff travel overseas for their holidays and become ill with gastroenteritis, they should visit a doctor, because there is a chance that they have acquired a foodborne illness.

It is the food business owner's responsibility to make sure their staff know about these requirements. It is preferable to do this in writing so that no points are missed; your EHO will be able to provide you with pamphlets or 'fact sheets' on this topic.

The level of risk associated with foodborne illness and food handling is highly variable, and depends on numerous factors, such as: the type of products being made, the nature of the illness or condition, and the duties of the food handler. Specific guidance should always be sought on a case-by-case basis from an EHO and/or relevant state or territory authority.

Staff who handle unpackaged potentially hazardous foods such as ready-to-eat or cook chill products must be vigilant in reporting if they have a confirmed or suspected case of foodborne illness. The food business may then need to take special measures; see Box 37 for an example.

It is important that you do not jump to conclusions if you see signs a staff member is suffering from a suspected foodborne illness. For example, a female staff member may vomit while at work

Box 36 – Foodborne illness: Conditions that may spread *Staphylococcus aureus*

Staphylococcus aureus is a normal part of the body flora for many people. If hygienic food handling practices are used the risk of contaminating food with *S. aureus* is decreased. However, under certain conditions there is an increased chance of transferring *S. aureus* into food during handling:

- Pus in skin infections, such as sores, boils, acne, cuts or abrasions, may contain high levels of *S. aureus*. If these infections are on the hands or parts of the body that may be touched by the hands (e.g. the face), there is a risk of transferring *S. aureus* onto food.
- Fluid coming out of an infected ear or eye may contain *S. aureus*, so if touched by the hands cells can be transferred onto food.
- Sneezing, coughing or blowing the nose when suffering from a cold, flu or allergic reaction is more likely to spread *S. aureus* from the nose or throat onto food.

If food handlers come to work with these conditions they should:

- completely cover areas of infected skin with a waterproof dressing, and wash hands thoroughly after changing the dressing (if this is required while at work)
- avoid touching infected skin lesions or fluid coming from the ear, nose or eye
- use tissues to blow the nose, and to sneeze or cough into, followed by immediate disposal of the tissue and washing hands thoroughly
- take medication to dry-up discharges caused by colds, flu or allergies.

Food handlers do not have to report they are suffering one of these conditions if:

- the infected skin is on parts of the body that are always covered with clothing and can't easily be touched
- cold, flu or allergy tablets are effectively 'drying' up any fluid coming from the nose.

Source: Standard 3.2.2 and Safe Food Australia

because she is pregnant, not ill. Additionally, you should consider that some staff members may be concerned that they could lose their job if they report they have contracted a foodborne illness. Because it is always best to err on the side of caution, it is recommended that you seek the guidance of your local EHO or food authority if you are unsure of the correct approach to take when a suspected case of foodborne illness in a staff member arises.

Box 37 – Packaged meat voluntary recall

In 2006, an Australian smallgoods manufacturer took appropriate precautionary measures when they voluntarily recalled ready-to-eat packaged meat. The manufacturer had learned that an employee had contracted hepatitis A. The staff member worked on the packing line for a range of shaved meats (ham, turkey, pastrami and chicken). Although there was no contamination of food reported, the recall and subsequent publicity raised awareness of the potential risk and alerted doctors to keep watch for symptoms.

Records related to staff foodborne illness

It is recommended that you keep records of foodborne illness suffered by your staff because these may be used if an EHO requests proof that you are meeting the requirements of the Code.

All information related to the health of individual staff members must be kept confidential unless permission is given by the affected staff member to disclose it. The only exceptions are if the information needs to be discussed with, or notified to, an EHO or another authorised representative of a state or territory authority (Standard 3.2.2).

Information that may be recorded includes:

- the date information was reported to you
- the symptoms (or illness if diagnosed)
- the actions taken (e.g. staff member stayed at home for 3 days)
- copies of any related doctors certificates provided.

Food hygiene essentials

It is a food business owner's responsibility to take all practicable measures to ensure no one on the food business premises contaminates food. This includes staff, visitors, tradespeople and contractors. The term 'practicable measures' takes into account that the actions of other people cannot always be controlled. For example, staff may choose not to wash their hands even though facilities are provided. However, there are many things that can be controlled, such as provision of properly maintained equipment and facilities (e.g. cold rooms operating at 5°C or below).

Responsibilities of food handlers

The Code (Standard 3.2.2) states that: 'A food handler must take all reasonable measures not to handle food or surfaces likely to come into contact with food in a way that is likely to compromise the safety and suitability of food.'

The term 'all reasonable measures' acknowledges that food handlers may not be able to meet this requirement if they encounter situations that are beyond their personal control. For example, if they have been given incorrect instructions from their supervisor, or if the equipment that is available for them to use is faulty. If the on-the-job or other form of food safety and hygiene training that food handlers receive does not adequately educate them on how to handle food safely, then this is also considered to be beyond their control. It is the responsibility of the owner of the business to ensure their staff receive appropriate training.

Specific steps food handlers can take to reduce food safety risks are provided throughout the remainder of this chapter; some other basic steps are to:

- report any issues of concern to their supervisor
- keep food covered when not handling it
- reduce the time potentially hazardous food is kept at room temperature
- make sure that cooking or processing instructions are followed
- follow cleaning and sanitising instructions
- ask for more information if they do not understand an instruction.

Hand washing

The requirements for providing hand washing facilities for staff have already been outlined earlier in this chapter.

These facilities must also:

- not be used for any other purpose other than washing hands – instruct staff not to use hand washing sinks for activities such as washing up dirty utensils, or washing vegetables (signs can be placed above hand washing sinks as a reminder)
- include a supply of soap – this can be any type of soap, including standard bar soap, liquid soap or antibacterial soap
- include a means to dry hands that is not likely to transfer pathogens to the hands:
  - disposable paper towels are the best option
  - non-disposable cloth towels may become contaminated and are not recommended
  - air dryers generally take much longer to dry hands fully compared with a paper towel, so there is a higher chance staff will not dry their hands properly
- have a waste container near the sink for disposal of used paper towels.

Box 38 – How to wash hands correctly

1. Wet hands

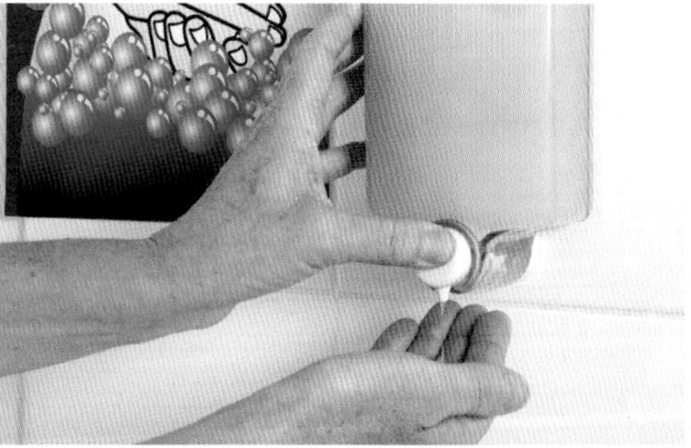

2. Apply soap

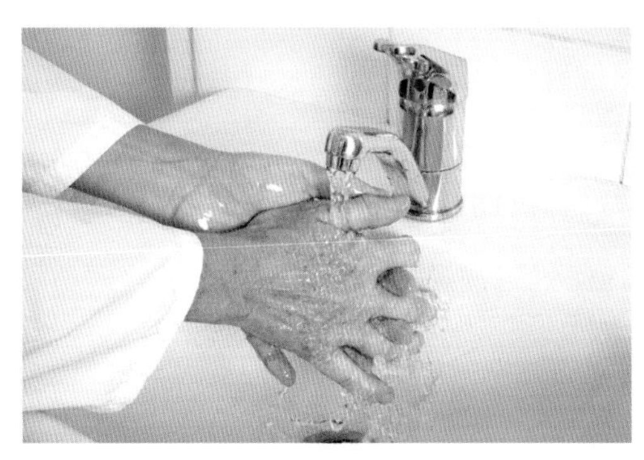

3. Wash hands thoroughly (including between fingers) for 20 seconds

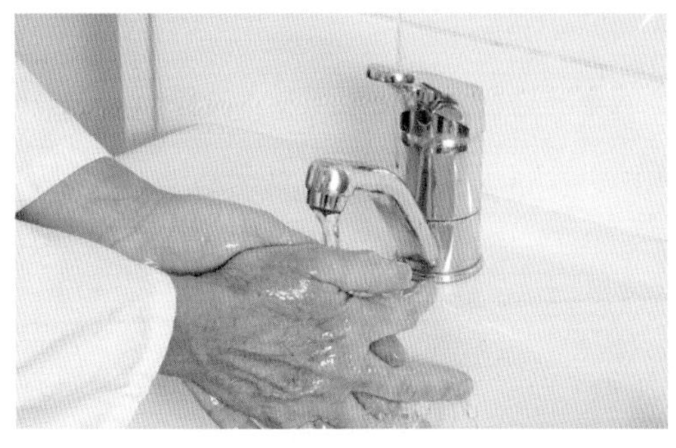

4. Rinse hands well

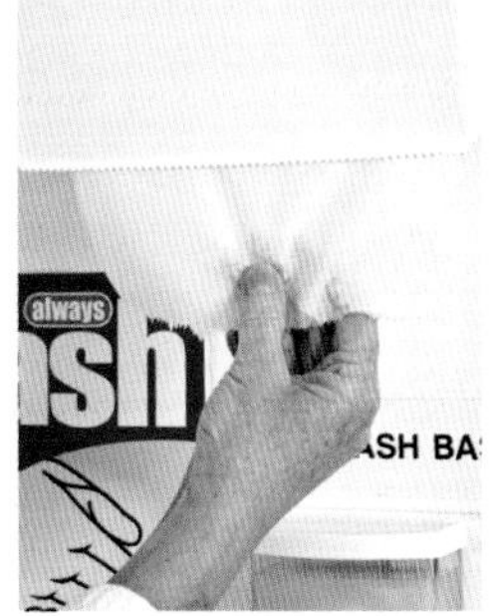

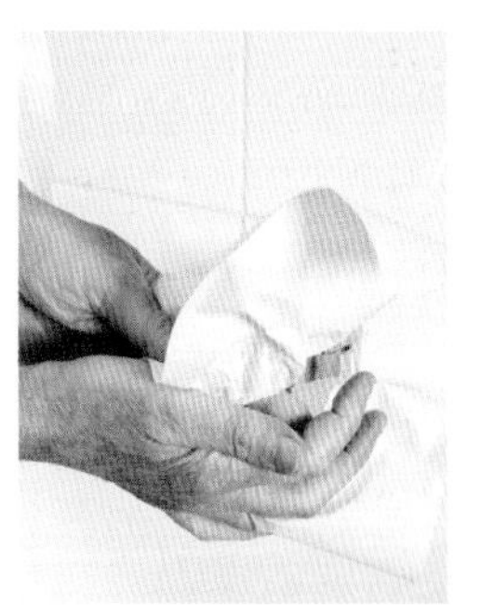

5. Dry hands completely

It is important that hands are dried thoroughly after washing because pathogens are more readily transferred onto wet hands (e.g. when touching surfaces such as door handles).

Staff must be trained in correct hand washing procedures and refresher training should be provided on a regular basis, with firm reminders provided if bad habits are observed. Box 38 shows the procedure recommended for effective hand washing.

Regular washing of hands reduces the risk of contaminating food, surfaces or equipment with pathogens that may be present on the hands.

Food handlers must wash their hands whenever they are likely to be a source of food contamination, such as:

- immediately before handling ready-to-eat food
- after handling raw food (particularly meat)
- immediately after using the toilet
- before directly touching food, equipment or surfaces likely to come into contact with food
- immediately after smoking, eating or drinking
- immediately after coughing, sneezing, or blowing the nose
- after touching the head, hair, mouth, nose or ears
- after handling garbage or cleaning
- after handling animals.

What staff can or cannot wear at work

It is important that you develop a policy on what clothing, uniforms and other items your staff can wear at work. This decision will be influenced by the level of risk associated with the duties of each staff member, and any requirements of your state or territory authority. Once you have determined what your policy is, you should inform your staff, preferably in writing so they can't say 'We weren't told that' if issues arise in the future.

Areas your policy can cover include:

- requirements for tying back long hair
- if hair covers are required and, if so, what type; requirements vary between states and territories, so check with an EHO
- if beard nets are required and, if so, when and where they should be worn
- cleanliness and acceptable length of fingernails (long nails are harder to clean under), and if polish or false nails are allowed
- if any jewellery or watches are permitted to be worn
- if staff wear their own clothes, guidelines that need to be followed; for example, bring clean ironed clothes from home and put them on once inside the premises

- if aprons or uniforms are provided:
  - areas where these may, or may not, be worn; for example, staff are not permitted to go 'down to the shops' during a break while wearing their uniform
  - what is the process for having these cleaned; for example, who is responsible and how frequently should they be cleaned
- if staff wash their own uniforms, guidelines on how this should be done.

Most food businesses do not allow their staff to wear jewellery and watches in food handling areas. There is a risk that whole items or pieces of them can fall into food. Additionally, pathogens or food particles can lodge under rings and make hand washing less effective. Staff may also wash their hands less thoroughly if they are trying to prevent their watch from getting wet. However, some businesses may permit staff to wear plain wedding rings or sleeper earrings. If you allow your staff to wear rings, it is preferable that they wear gloves if they handle food directly and the rings should be plain to reduce the chance of the ring catching on gloves and making a hole.

If your policy states that you will provide uniforms for staff, it is your responsibility to always have an adequate supply of clean items available in an appropriate range of sizes. It is preferable that uniforms and aprons are lightly coloured because it is easier to see when they become dirty. Different areas can have different coloured uniforms to make it easy to identify if someone who should not be in an area is in the wrong place.

If clothing or uniforms routinely become dirty during the day because of the type of tasks performed, it is good practice to supply aprons. Aprons should be changed if swapping from working with raw foods such as uncooked meat, to ready-to-eat foods such as cooked meat or salad items. Aprons should not be worn when using the toilet or going outside.

Sleeves on clothing or uniforms should not be so long that they can 'dangle' in food, which could cause cross contamination from one food to another. It is preferable that ties or Velcro™ fasteners are used instead of buttons or press-studs. Any loose buttons or studs should be repaired because there is a risk of them falling into food and becoming a physical hazard.

Covering wounds and sores

Although you may not generally require your staff to wear gloves, if they have a wound on their hand that needs covering, the most effective way of doing this is by wearing a glove over an adhesive dressing or bandage. It is a requirement of the Code that waterproof wound coverings

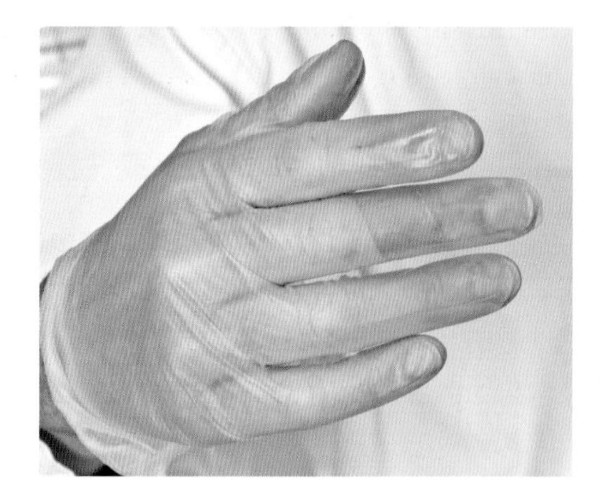

are used in food handling areas. This reduces the chance of the dressing falling off, which can occur more readily if it gets wet, or if blood or pus seep out from under the dressing. It also reduces the risk of the wound dressing falling off and contaminating food.

You should provide a supply of wound dressings for staff to use. Blue adhesive dressings are used most commonly because they are easier to see if they fall into food. Alternatively, you can use brightly coloured children's plasters that are available from supermarkets.

Personal habits and cleanliness

Staff who handle food, or any items that may come into contact with food, should maintain a good level of personal hygiene, particularly if dealing with ready-to-eat foods. Importantly, hands and fingernails should be kept clean at all times. It is likely that occasionally throughout the day staff will touch their hair or clothing with their hands: potentially transferring pathogens onto the hands. To minimise this risk, it is better if hair and personal clothing is clean when staff start work. If hair is kept covered at work, or if staff change into uniforms, less emphasis can be placed on this requirement.

Although it is difficult because these actions are usually done unconsciously (i.e. not deliberately), staff should try and avoid habits such as touching their nose, scratching their skin, chewing their fingernails or playing with their hair while handling food or food contact surfaces. These and other similar actions increase the risk of hands becoming contaminated with pathogens that can be transferred onto food.

Food handlers should avoid getting any of their body fluids onto food. Covering the mouth when sneezing or coughing and using medication to dry-up runny noses reduce the risk. Hands should always be washed after covering the mouth for a sneeze or cough, regardless of how this is done (e.g. hands, tissue or handkerchief). Similarly, if staff blow their nose in the middle of handling food, they must wash their hands before recommencing.

Eating food or chewing gum must not be permitted in any food handling area. Pieces of food or gum can fall onto food, equipment or surfaces. This can cause physical contamination or the transfer of pathogens. Eating food with the hands also increases the chance of fingers becoming contaminated with pathogens in saliva or on the skin around the mouth. The only exception to this requirement is if a staff member needs to taste the food being prepared to check if seasonings need adjusting (as is the common practice in restaurants). Disposable spoons should be used for this purpose and they should be discarded immediately and not re-used after being in the mouth.

Staff must report possible food contamination

Although staff should have been provided with adequate food hygiene training and the equipment and facilities needed to handle food safely, accidents can still occur. The Code specifies that food handlers must tell their supervisor if they know or suspect they may have contaminated food during handling.

Examples of different scenarios requiring reporting include:

- a chopping board used to cut-up raw chicken was used to chop salad vegetables without being cleaned and sanitised in-between
- a screw cap from a container of cooking oil has been misplaced and cannot be found
- an adhesive dressing used to cover a wound on a finger has fallen into a bowl of food
- an uncovered container of shredded lettuce was left on the bench when a high-pressure hose was used for cleaning dirty equipment nearby.

If a staff member does report that they may have contaminated food, you should bear in mind that the way you react may affect the likelihood of staff reporting any incidents to you in the future. If a staff member repeatedly makes mistakes or forgets things that affect food safety, further training and/or closer supervision may be required.

Extra precautions when handling high-risk foods

Although all food must be protected from contamination, foods that are higher risk are those that will not be heated at all (i.e. ready-to-eat) or will only be re-heated by consumers (e.g. cook chill). The strictest of hygienic practices are required when handling these foods.

Box 39 – Correct use of gloves

If not used correctly, gloves can present a larger food safety risk than if they were not used at all. Staff can mistakenly think that because they are wearing gloves they do not need to wash their hands as frequently. This is not correct, because gloves get contaminated in just the same way that hands do.

Some tips for safe use of gloves are:

- gloves should be disposable and not re-used after removal
- hands should be washed and dried thoroughly before putting gloves on
- gloves should be tough enough for the types of tasks performed
- gloves must be changed between handling raw and ready-to-eat food
- when putting gloves on, the glove surfaces that will come into contact with food should not be touched with bare hands
- gloved hands should be washed, or the gloves changed, if glove contamination is suspected.

If you do require your staff to wear gloves for certain food handling tasks, you need to provide specific information about how and when to wear gloves and how frequently they must be changed.

Extra precautions are required between cooking or other processing used to reduce the levels of pathogens (e.g. sanitising vegetables), and sealing in the final packaging; otherwise you risk post-process contamination with pathogens, some of which only need to be present at low levels to cause illness, such as hepatitis A.

Staff education should include instructions on the importance of separating work activities, when duties involve handling both raw and ready-to-eat or cooked foods. Production should flow progressively from delivery of raw materials through to distribution of finished products. Backtracking or crossovers of people, food, utensils or equipment 'against the flow' should be avoided.

If you have very limited space, you may only be able to provide this separation by controlling the time when tasks are performed, allowing a single area to be used. Let's look at the example of a business with only one bench, making a variety of different meat-based casseroles. Instead of chopping up one batch of meat then preparing a casserole, followed by chopping up another batch of meat and preparing the next casserole, all the raw meat should be prepared in the morning and then stored in a fridge or cold room until required later in the day. After the raw meat handling steps are completed, the bench top, utensils and other items used must be cleaned and sanitised before re-use. Alternatively, the business could prepare the raw meat at the end of the day ready for cooking the following day.

Other general steps that can be taken to reduce the chance of contamination of higher risk foods include:

- providing commercially laundered uniforms and aprons to all staff
- reducing the need for staff to handle food with their bare hands by providing items such as:
  - utensils, such as tongs, spoons and spatulas
  - gloves, which must be used correctly (Box 39)
  - other barriers, such as plastic bags or paper towels
- if food does need to be touched directly with the hands, it is important they are washed thoroughly beforehand.

If you have room on your premises, you should consider setting up separate designated areas where high-risk foods can be handled after they are cooked, processed or sanitised.

Specific procedures are recommended for these 'clean areas':

- restrict entry to clean areas only to staff trained to handle these products
- require staff to change their apron before entering these areas, particularly if they have been handling raw, unprocessed or un-sanitised food in other areas. Use different coloured aprons in these areas, so staff who have not changed their apron can easily be spotted.
- require staff wash their hands before entering these areas
- require staff who work in these areas to wear a hair net
- consider installing footbaths containing sanitiser at the entry point to these areas; if using these, you will need to provide your staff with appropriate footwear otherwise they won't use the footbath for fear of damaging their shoes; additionally, the sanitiser needs to be changed frequently.

Controlling the activities of visitors

Many small businesses are family orientated; this means there is a chance of children, relatives and family friends visiting your premises. Other potential visitors include cleaners, pest-control contractors, tradespeople, delivery people and salespeople. These people should either be supervised at all times or given clear instructions on which areas they are permitted in and what they can or can't touch. Children must be closely supervised at all times. It is preferable to arrange for tradespeople and cleaners to come to your premises during times when there is no food preparation or handling occurring.

KEY MESSAGES FROM CHAPTER 3

- All food businesses must be registered with their local council or the appropriate authority in their state or territory.
- The layout of premises must minimise opportunities for food contamination (e.g. raw foods separated from processed foods).
- Essential services and equipment must be available, such as hand washing facilities and adequate chilled storage space.
- Potentially hazardous foods may contain pathogenic microorganisms and are able to support growth or toxin production by these pathogens.
- Potentially hazardous foods must be stored at 5°C or below, at 60°C or above, or at another temperature for a limited time if it can be shown that this is safe. This is a requirement of the Code.
- An effective cleaning and sanitation program must be in place to maintain an adequate level of hygiene in food handling or processing environments. This is a requirement of the Code.
- An effective pest-control program must be in place to prevent pests bringing contaminants into food business premises. This is a requirement of the Code.
- Overall accountability for the safety of products sold by a business lies with the owner of the business; it is their responsibility to ensure no one on their premises contaminates food.
- Food businesses must ensure food handlers have food safety and hygiene skills and knowledge relevant to their work duties. This is a requirement of the Code.
- Food businesses may be required to have a Food Safety Supervisor, who is certified by a Registered Training Organisation.
- Food handlers experiencing symptoms of a foodborne illness or known to be suffering from a foodborne illness must not handle food. They may only re-commence handling food after the food business has received advice from a doctor that they are no longer suffering from or carrying a foodborne illness. These are requirements of the Code.
- Extra food hygiene precautions should be used when handling high-risk foods, such as ready-to-eat or cook chill products.

Product: Hummus

Formulation
Chick peas
Water
Tahini
Lemon juice
Garlic, crushed
Salt
Preparation:

Ingredients:

Tomato pulp (1.045 Specific Gr
10% acetic acid vinegar
Sugar
Salt

Barbara's Butters

LEMON BUTTER	
Requirements:	Preparation:
22.5 Kg sugar	1 Add ingredients, lemon oil and star bring to the boil.
9 Kg glucose syrup	2 Mix starch with j add, along with le
3 Kg margarine	3 Bring back to the 5 minutes.
pectin	4 Take a sample, co and measure and r
	5 Adjust the pH wit ary to bring

dave's delish dips

Baba Ghannous

Ingredients:
Eggplant
Lemon juice
Tahini
Garlic paste
Olive oil
Salt

Procedure:

Preheat oven on sett
Roast eggplant for 3
Remove and skin e
Puree flesh while h
tahini.
Add crushed gar
blending.
Remove sampl
Package if pH
Call Toni if

MARVELLOUS MAYONNAISES

Basic formulation

Vegetable oil	80 kg
Pasteurised egg yolk	8 kg
Water	6 kg
Vinegar (10% acetic acid)	2 kg
Sugar	1 kg
Salt	1 kg

Blend egg yolk, water, vinegar, sugar and salt until all is dissolved.
Slowly blend in oil to form an emulsion.
Check pH and record.
Adjust if necessary.

Chapter 4

Controlling food safety hazards – your product recipes

You have already learnt that pathogenic microorganisms can be controlled by altering product recipes. Recipe hurdles – which either singly or in combination with other factors such as refrigeration, result in safer products – include adding acid, reducing the amount of water that is available in the product and adding preservatives. This chapter provides practical guidance for how to set up these recipe hurdles.

In some cases, what you learn here may make you realise you need to 'tweak' a recipe to make it safer. If your recipe can't be adapted sufficiently to ensure the safety of your product right through to when it is in the hands of the consumer, perhaps the recipe should be discarded. In other cases, you may discover that your product already provides the appropriate hurdles without any need to change the recipe.

Food products that rely heavily on the following controls will not be discussed in this chapter:

- heat processing and packaging techniques that achieve 'commercial sterility' (i.e. traditional canning)
- frozen storage – correct frozen storage prevents growth of all microorganisms
- dried and semi-dried foods (less than 25% water) – low available water content inhibits growth of pathogens.

Adding acid to food

Adding acid to food lowers the pH, which is a measure of acidity. pH is measured on a scale of 0 to 14: a pH of 7 is neutral; a pH above 7 is alkaline; and a pH below 7 is acidic. The lower the pH

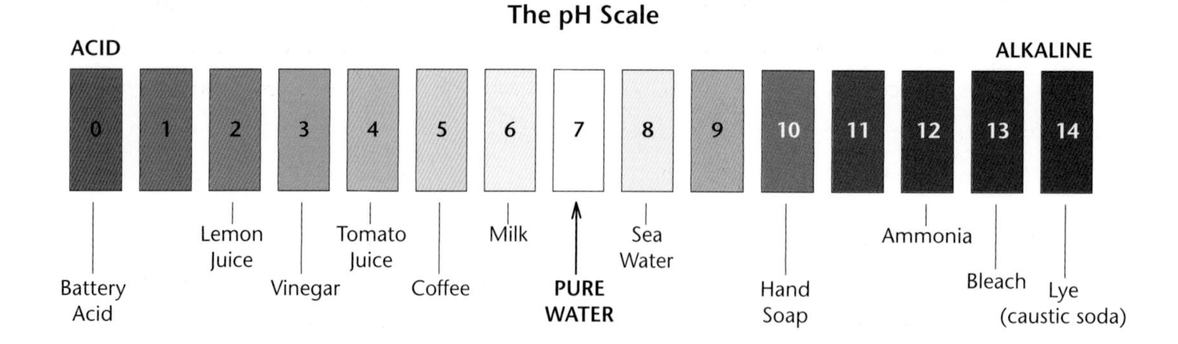

of the food, the safer it will be, because pathogens find it harder to grow and dormant bacterial spores don't readily become activated in acidic environments.

Very few foods have a pH above 7 because most foods naturally contain some acid. 'Low-acid foods' have a pH above 4.6. 'High-acid foods' have a pH below 4.6 (to see why pH 4.6 is so significant see Box 40).

The natural acidity level of foods

High-acid foods (below pH 4.6)	Foods with a pH about 4.6	Low-acid foods (above pH 4.6)
Apples	Bananas	Red meat
Apricots	Figs	Poultry
Lemons	Mangos	Seafood
Oranges	Pawpaws	Vegetables
Pineapples	Tomatoes	Garlic
Raspberries		Black olives

Foods that may require additional acid to reduce the risk of the growth of pathogens include chilled fresh dips, pasta sauces, mayonnaise and dressings such as aioli. These foods are generally in the low-acid category, or have a pH close to 4.6 and need to have their pH reduced to below 4.6 by the addition of extra acid.

It is a requirement of the Code that all fruit and vegetables in brine, oil, vinegar or water must not have a pH above 4.6 (Standard 2.3.1 Fruit and Vegetables). This means they must have a final pH of 4.6 or below (high acid). Canned foods are exempt from this regulation. See Box 41 for details of how to acidify vegetables or herbs that will be stored in oil.

Different acids permitted for use in Australian foods include acetic, citric, lactic, malic and tartaric (the Code, Part 1.3 Substances Added to Food). The two acids used most often by small food businesses are acetic acid (in the form of vinegar) and citric acid (either from lemon juice or in powder form). The other permitted acids are used more rarely because they are quite expensive by comparison and their flavours are not always compatible with the product.

Different forms of acetic and citric acid are available and the concentration of acid contained in each type varies. For example, white vinegar purchased from a supermarket generally contains about 4% acetic acid, while larger volume vinegars available for food manufacturing use contain more (e.g. 10% acetic acid).

You can choose to use either acetic or citric acid in your product, or a combination of both. Citric acid lowers the pH more effectively than acetic, but acetic acid has better preservative action. Your choice of which acid to use will depend on the type of product you are making and how well the flavour of the acid fits with the flavour of the product. The flavour and aroma of acetic acid are noticeably stronger than citric acid. The best way to adjust the pH of your product and still be satisfied with its flavour is a trial-and-error approach. You may also wish to experiment using different levels of sugar in your product, which may 'balance out' the tartness of the acid.

Avoid making too many assumptions regarding the pH of your product. You may think that because the main ingredient is acidic, the overall pH will be below 4.6 (high acid); this may not always be the case and it is always safer to measure the pH of your final product. See Box 42 for an example.

There are two ways to measure the pH of food: pH test strips or a pH meter.

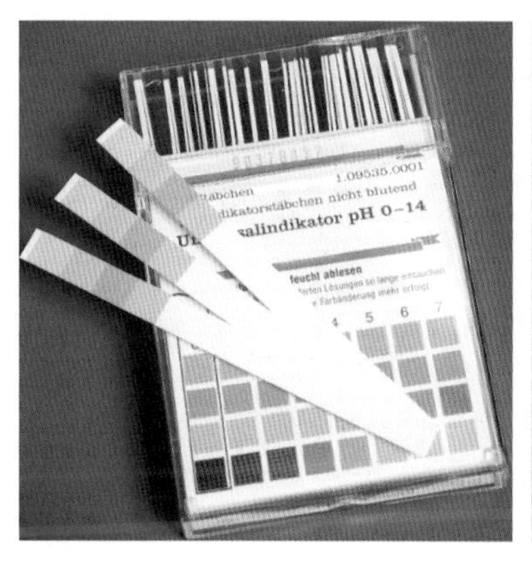

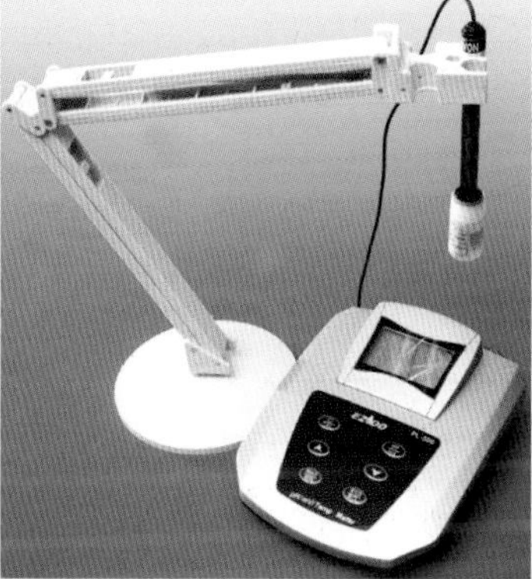

Box 40 – pH 4.6 and food safety

Clostridium botulinum, a dangerous pathogen, produces heat resistant spores and vegetative cells that are unable to become activated or grow below pH 4.6.

If the pH of your product is above pH 4.6 (low acid), there is a risk of *C. botulinum* cells growing in your product and producing a deadly toxin.

Foods in this category must either be:

- stored chilled strictly at 5°C or below to slow cell growth, but only for a restricted time (depending on what other hurdles have been used)
- stored frozen to prevent cell growth
- processed at temperatures above 100°C for a predetermined time, in air and watertight packaging, to kill any *C. botulinum* present and stop it from re-contaminating the product (i.e. achieving commercial sterility). This type of processing must not be performed without first seeking expert professional advice (see Chapter 6 for more information).

The last method is how low-acid, shelf-stable canned foods are prepared and the minimum process required is commonly called a 'botulinum cook'. Most people preparing food on a small-scale will not have access to the specialised equipment required to adequately achieve a full botulinum cook.

The use of lower heating temperatures can achieve safe shelf-stable products, but first the pH of all components of the product must be reduced below pH 4.6 (making it high acid). One example of the use of this technique is the process of pickling vegetables, which is discussed in Chapter 6 (Box 63).

Box 41 – Preservation of vegetables and herbs in oil (unpasteurised)

Many small businesses in Australia make unpasteurised vegetables bottled in oil and herb-infused flavoured oils. The types of products made include chopped garlic, whole garlic cloves, sun-dried tomatoes, olives, chillies, ginger, roasted eggplant, roasted capsicum, mushrooms and herb-infused oils.

Unless correctly prepared, these products pose a real risk of foodborne illness. This was demonstrated by two 1980s outbreaks of botulism – a serious foodborne illness caused by *Clostridium botulinum* toxin – in the United States and Canada. Because *C. botulinum* can grow in environments in which there is no oxygen, suspending low-acid foods (above pH 4.6) in oil creates an ideal environment for this to occur.

To ensure the safety of these products, the Code specifies that they must not have a pH above 4.6 (Standard 2.3.1).

The correct way to safely prepare these products is to add vinegar to the vegetables before adding the oil. If standard domestic vinegar containing 4% acetic acid is used, at least one-quarter of the total weight of vegetable or herbs (not including the oil) should be as vinegar. For example, to make a 400 g garlic and vinegar mix, use 300 g garlic and 100 g vinegar. Once the vinegar has been added allow the mixture to stand for 24–48 hours then check the pH is well below 4.6 using the method described on page 119. Once the pH is below 4.6, the oil can be added.

To check the pH after bottling, you should remove the vegetables, blot off as much oil as possible with paper towel and then puree.

Once you have worked out how much acid to add to how many vegetables or herbs, you must always weigh out the correct quantity of each and always use the same concentration of acid. After adding the acid and before adding the oil, you also need to measure and record the pH of every product batch. This way you have a record that the acid was correctly added to each batch.

In addition to this acidification step, you may need to refrigerate these products to prevent the growth of spoilage microorganisms such as yeasts and moulds.

Box 42 – Dips made with a yoghurt base

Because yoghurt is a high-acid food with a pH about 4, it is unable to support the growth of pathogenic microorganisms. What happens though when you start to add low-acid foods to yoghurt? This dilutes the acid in the yoghurt, resulting in a product with a final pH higher than the original yoghurt.

Take, for example, the traditional Greek dip, tzatziki. The addition of cucumber and garlic to the yoghurt base may bring the pH of the final product close to 4.6, perhaps even above 4.6 (making it low acid). It would then be necessary to add some acid to the dip until the final pH was well below 4.6 (high acid).

Box 43 – Products containing cooked or raw meat (whole-cuts, pieces or minced)

All meats and flesh foods are low-acid foods, with a pH above 4.6. Attempting to lower the pH of these to below pH 4.6 can be very difficult to achieve because acid generally does not penetrate very well into meat. This makes meat harder to acidify than other foods, such as vegetables.

If your product contains a mixture of meat and other ingredients, say as a sauce, you must measure the pH of the meat separately from the sauce. Rinse any sauce off the meat with deionised water, weigh the meat and then add an equal weight of deionised water before pureeing. You must not puree the meat and sauce together before measuring the pH. This may give you false confidence in your product's safety if this results in an overall pH below 4.6. Although the growth of pathogens will be restricted in the sauce, it will not be restricted within the meat component.

Important note – in general, products containing meat must not rely on a reduced pH as the sole food safety control because the pH can't be reliably adjusted to below 4.6.

Unless your product is below pH 4, a pH meter should be used because it offers a higher level of accuracy (see page 277 for information on where to find suppliers of pH meters). Before measuring the pH of your product, it is essential that it is in liquid form and that solid pieces are dispersed evenly; this is best achieved by pureeing. Use one part of deionised water to one part of solid to puree; that is, half water and half food. Measure the pH of the resulting mix. If you notice that the reading changes when you move the pH meter probe around, you may need to puree the sample more or wait until the reading stabilises. The length of time required will depend on the type of product you are testing.

Foods that have been preserved in oil and vinegar, and foods containing ingredients with a high level of protein (e.g. meat), are special cases. To learn more about measuring the pH of these see Boxes 41 and 43.

Reducing the available water in food

As outlined in Chapter 2, microorganisms require a certain amount of available moisture in food to survive and grow. Controlling the amount of moisture available in your product can restrict microbial growth.

The term 'water activity' is used when describing the water availability in food, and you may see it abbreviated to a_w. The water activity scale extends from zero (bone dry) to one (pure water). Water activity does not simply mean the amount of water you use in a recipe: it relates to the water-binding ability of the ingredients used. If water is bound within ingredients, it is not available to support the growth of microorganisms. Bacteria, yeast and moulds either can't grow at all in low-water-activity products, or they can grow only very slowly.

Examples of products with low water activities are jam and anchovies. Jam contains about 40% water but because jam has a high amount of sugar dissolved into the water, this water is 'tied up' or bound. The water in anchovies, on the other hand, is bound by a high salt concentration.

Some examples of different food types and their water activities are shown in the table below.

Water activities of some foods

Food	Water activity
Fresh meat	0.99
Bread	0.95
Aged cheddar cheese	0.85
Jam	0.7
Anchovy fillets	0.7
Dried fruit	0.6
Dry biscuits	0.3
Instant coffee	0.2

Just as *Clostridium botulinum* has a cut-off point for growth at pH 4.6, there are certain cut-off points associated with water activity and the growth of pathogens (Box 44). Foods can be made safer by reducing the water activity to below the level at which pathogens can grow.

Techniques used to reduce water activity include:

- removing water – for example, drying or evaporating
- adding ingredients that will bind to water – for example, adding curing agents and/or salt to meat or adding sugar to make jam.

Box 44 – Water activity and food safety

Products stored at room temperature:

Staphylococcus aureus is able to grow and produce toxin in products with a water activity above 0.85 and a storage temperature of 10°C or above. The Code, Standard 1.6.2 Processing Requirements, specifies that the water activity of dried meats must not be higher than 0.85 – this is to prevent growth and toxin production by *S. aureus* in these types of products, which are generally stored at room temperature (e.g. beef jerky).

Products stored under refrigeration:

There is one type of *Clostridium botulinum* that is able to grow and produce toxin at temperatures as low as 3°C in foods above pH 4.6. However, it is only able to do this in foods with a water activity above 0.97. Because it is not practical to store most refrigerated products below 3°C throughout their entire storage and distribution cycle, reducing the water activity of products to 0.97 or below is one way to control the growth of this type of *C. botulinum*. Storage of food at 5°C or below will control the growth of the other type of *C. botulinum*.

Salt and sugar are commonly used to reduce water activity, and different forms may be used to suit your particular product; for example, using sugar syrup instead of sugar granules. Different sugars have different abilities to control water activity. Glucose (dextrose) is more effective than sucrose (cane sugar), which is more effective than some maltodextrins (derived from starch). Sugars also have varying levels of sweetness. For example, glucose is only about 60% as sweet as sucrose; fructose is the sweetest sugar, and is estimated to be twice as sweet as sucrose. Glycerol (or glycerine), a sweet tasting liquid, is very effective in lowering water activity.

Use of appropriate packaging will assist you to maintain the desired water activity in your product. This is particularly important if you use fructose to control water activity because it tends to absorb moisture from the air. If you use fructose then it must be stored under moisture-proof conditions. Advice for choosing appropriate packaging types will be provided in Chapter 7.

Information on how to find suppliers of equipment or kits to measure water activity can be found at the end of the book. Alternatively you can arrange for a laboratory to test your products, particularly if you are in the product development stage.

When to measure pH and water activity

Before going out and buying a pH and water activity meter, you need to consider if you need to measure these values. The following table gives you some guidance for different product types.

If you have determined that your products pH or water activity is a food safety hurdle, you must not make any changes to your recipe that may alter these without first re-measuring to check the new values are still within the 'safe range'. See Box 45, for an example of what can happen if this is not done.

Box 45 – When things go wrong: changing product recipes

In England during 1989, there was an outbreak of the foodborne illness botulism, which is caused by *Clostridium botulinum* toxin. Twenty-five people were affected and one person died. Hazelnut puree used to flavour yoghurt caused the outbreak.

The puree normally contained enough sugar to keep the water activity at a level where *C. botulinum* could not grow. This meant just a light pasteurisation heat process was required for product safety. However, the manufacturer decided to make a low sugar version and replaced the sugar with an artificial sweetener. This raised the water activity significantly, to a level where *C. botulinum* could grow.

The puree now required a botulinum cook to control the growth of *C. botulinum*. However, because the manufacturer did not consider what effect changing the recipe would have on the safety of the product, it continued to use a light pasteurisation heat process. Consequently, *C. botulinum* grew and produced toxin in the puree.

Deciding when to measure hurdles

Product characteristics	Example	Measurements	
		pH	Water activity
Moist, containing low-acid (above pH 4.6) ingredients	Vegetable dip	Yes	No
Moist, primarily containing high-acid (below pH 4.6) ingredients	Lemon butter	No	No
Dry, baked goods	Bread	No	No
When the Code stipulates	Vegetables in oil	Yes	No
	Dried meats	No	Yes
When specifically required to control pathogens	Some fermented meats	Yes	Yes

Adding chemical preservatives

Chemical preservatives are food additives that extend the shelf-life of food by protecting against deterioration caused by microorganisms. Some preservatives also play a role in enhancing food safety, and these are the focus of this segment.

There are several chemical preservatives permitted for use in foods sold in Australia; these are specified in the Code, Part 1.3 Substances Added to Foods. Addition of any other chemical preservative is not allowed. The Code specifies in which foods each preservative is permitted and, in most cases, the maximum concentration you are allowed to add. In some cases, instead of a concentration, the Code states the level of a preservative permitted in a specific food type must be in accordance with Good Manufacturing Practice (GMP); see Box 46 for more details.

Nitrites and nitrates

The use of nitrites and nitrates (potassium and sodium salts) is permitted in some types of smallgoods to stop the activation and growth of certain spore-forming pathogenic bacteria, notably *Clostridium botulinum*.

Cured meat and fish products receive a relatively mild heat process, which reduces the number of bacterial cells in the product but has little effect on heat resistant spores. Preservative use, coupled with either storage at 5°C or below or drying to reduce the water activity, is essential to maintain the safety of these products.

In their Guidelines for the Safe Manufacture of Smallgoods (see page 272), Meat & Livestock Australia provide specific recommendations for nitrite and nitrate use in smallgoods. The types of smallgoods discussed include uncooked and cooked fermented meats, cured meats and fish, and some speciality products such as pâté. The Guidelines are essential reading for anybody intending to produce smallgoods.

Nitrites and nitrates are relatively toxic chemicals and great care is required to make sure permitted levels are never exceeded. Different maximum levels are specified for different product categories; refer to the Code (Standard 1.3.1) for details.

Occasionally incidents occur where a mistake in addition of curing salts has resulted in consumer illness. Curing pre-mixes contain instructions for correct addition of curing salts and these should be strictly followed. If you are weighing out your own preservatives, you must strictly control the procedure so that errors do not occur.

Nisin

Nisin is another preservative that inhibits activation of bacterial spores, preventing them from changing into cells and growing. It is mainly used in the cheese industry to prevent spoilage caused by late 'blowing' of cheeses owing to the growth of spore-forming bacteria.

Nisin is also used to control bacterial spores in high-moisture bakery products including crumpets and pikelets. These products have occasionally been implicated in foodborne disease outbreaks caused by *Bacillus cereus*, which is able to survive the mild baking processes involved.

Some other product categories in which nisin is permitted are:

- tomato products (with a pH below 4.5)
- fruit and vegetable preparations (including pulp)
- processed meat, poultry and game products in whole cuts or pieces
- dairy- and fat-based desserts, dips and snacks
- sauces and toppings.

Box 46 – Good Manufacturing Practice in the use of additives (including preservatives)

The following criteria are used to assess compliance with GMP relating to the addition of preservatives and other food additives:

- The quantity of additive added to food should be limited to the lowest possible level necessary to accomplish its desired effect.
- The quantity of the additive that becomes a part of food as a result of its use in the manufacture, processing or packaging steps and which is not intended to accomplish any physical or other technical effect in the finished food itself, is reduced as much as possible.
- The additive is prepared and handled in the same way as a food ingredient.

Claims made on product labels may affect the type and level of food additives that could be used in accordance with GMP; for example, the use of terms such as 'natural', 'pure' or 'traditional'. Similarly, the type and level of food additives used may affect the way in which food can be labelled.

Source: Standard 1.3.1 Food Additives

Nisin is allowed to be used in permitted products in accordance with GMP, and does not have a maximum permitted concentration specified.

Sulphur dioxide and sulphites

Sulphur dioxide and sulphites (sodium and potassium) are permitted for use in a range of foods to control growth of spoilage microorganisms, such as yeasts and moulds in fruit and vegetable juices. They are also permitted in some other products, most notably sausage mince, where they are effective in extending shelf-life and controlling growth of some pathogenic bacteria.

Sulphur dioxide can cause severe reactions in sensitive people, and can trigger asthma attacks in susceptible individuals. It is a requirement of the Code that the presence of more than 10 mg per kg of food be stated on the label to warn sensitive consumers of its presence. This is necessary even when it is present in your product only because it was added into one of your raw materials by another manufacturer. Concentration limits specified in the Code must be strictly followed and precautions used to make sure the correct amounts are added to every product batch. Although sulphur dioxide is driven off when foods are heated, some will remain bound to the food and still be active in the final product. You may need to get a laboratory to test your final products for sulphur dioxide content.

Sulphur dioxide is permitted in sausage mince, but it is not permitted in minced or ground meat. There is a history of sulphur dioxide being added illegally to minced meat because it helps maintain the fresh red colour of the mince as it ages. This can pose a health hazard to those people who are sensitive to this chemical but who would not expect minced meat to contain it.

Combining hurdles to control pathogenic microorganisms

So far, you have learnt about restricting the growth of pathogens by controlling temperature, reducing pH, reducing water activity or adding preservatives. For some products, these recipe hurdles are not used in isolation, but one or more combinations is used.

This can either:

- enhance the overall food safety and quality
- give greater flexibility when formulating recipes containing low-acid ingredients, where the addition of too much acid could damage the flavour
- provide an extra safety margin for products with a 'borderline' water activity or pH (e.g. pH 4.5).

Refrigeration at 5°C or below coupled with restricted shelf-life is a major hurdle used to control the growth of pathogens. However, as many of the foodborne pathogenic bacteria discussed in this book are able to grow at 12°C or below, it is better not to rely on refrigeration alone to control the growth of pathogens for any longer than a few days.

Some, or even all, of your products are likely to be stored above 5°C for some period of time after leaving your premises, even if they are labelled 'Store at 5°C or below'. This 'temperature abuse' can occur any time in the products storage and distribution life, but is even more likely to happen after it is purchased by the consumer. Results of a 1997 Australian survey of the general public showed that only 26% of those surveyed knew what the correct temperature of their fridge should be (i.e. 5°C or below). This is why it is essential to use additional hurdles, not just refrigerated storage.

Examples of products that rely on refrigeration in combination with other hurdles are listed in the table below.

Combining food safety hurdles

Food	Hurdles
Dips	Reduced pH and refrigeration
Pâté (cooked)	Low water activity on surface of pâté, cooking and refrigeration
Vacuum packaged cured meats	Preservatives, refrigeration and sometimes salt
Fermented meats (uncooked), (see Box 47)	Reduced pH, low water activity, preservatives and refrigeration

Box 47 – Uncooked fermented meats and food safety

Uncooked comminuted fermented meat (UCFM) products include smallgoods such as mettwurst, bologna, mould-ripened and Italian salamis. In the Code (Standard 4.2.3), UCFM is defined as, '… comminuted fermented meat which has not had its core temperature maintained at 65°C for at least 10 minutes or an equivalent combination of time and higher temperature during production'. Therefore, products that have received a mild heat process may still fall into the UCFM category.

UCFM products can be a serious and real risk to consumers if adequate food safety controls are not used. In 1994–1995, an *E. coli* foodborne illness outbreak linked to mettwurst manufactured in South Australia occurred. Approximately 150 people became ill, 23 children were hospitalised, and one child died.

Those involved in the production of UCFM products must be certain they have adequate food safety controls to prevent these products from causing foodborne illness. The Code (Standard 4.2.3) states specific requirements that must be met by those manufacturing UCFM products, including the need to have an effective food safety management system.

Additionally, Meat & Livestock Australia provide specific guidance in their Guidelines for the Safe Manufacture of Smallgoods.

Additional hurdles should always be used when the final pH is close to pH 4.6 to provide a sufficient margin of safety.

Altering your recipe by adding an extra ingredient, such as acid, may result in a better tasting product. For example, you can maintain the fresh flavour in a tomato-based sauce by reducing the cooking time, but maintain safety by adding extra acid to reduce the pH.

KEY MESSAGES FROM CHAPTER 4

- Low pH, acidic foods are safer, because pathogens find it harder to grow or become activated in acidic environments.
- *Clostridium botulinum* is unable to grow and produce toxin in foods below pH 4.6 (high acid).
- All fruit and vegetables in brine, oil, vinegar or water must have a pH below 4.6. This is a requirement of the Code.
- Reducing the amount of moisture available in food can restrict the growth of pathogens.
- The term used to describe the available water content of foods is 'water activity'.
- Salt and sugar are commonly added to foods to reduce their water activity; drying food also reduces water activity.
- The Code specifies the types of chemical preservatives allowed in different foods and at what level they may be added; addition of any other chemical preservatives is not permitted.
- Combining several food safety control measures or 'hurdles' enhances the safety of foods; combinations include:
  - reduced pH and chilling at 5°C or below (for a limited time)
  - reduced water activity, heating, chilling at 5°C or below (for a limited time)
  - chemical preservatives, chilling at 5°C or below (for a limited time).
- Chilling should not be relied upon as the only control measure for pathogenic microorganisms, because some can grow slowly under refrigeration, and maintaining temperatures of 5°C or below can't be guaranteed during the whole shelf-life of the product.

Chapter 5

Controlling food safety hazards – your ingredients

Most of you will recognise that to make a high-quality product you need high-quality raw ingredients. This also applies to safety; to produce safe foods you need to start by using safe ingredients. The first step is to have checks in place to monitor the safety of the ingredients you purchase. Then you need to transport, receive and store these ingredients in a safe manner.

You must consider product safety when deciding on the specific types of ingredients you use; for example, using pasteurised egg products rather than fresh eggs in products that are not cooked or are only lightly cooked, such as tiramisu (to see why, read Box 48). Also, whenever possible, you should substitute ingredients that are food allergens with those that are non-allergenic (e.g. thicken a sauce with maize starch rather than wheat starch).

However, the safety of ingredients also depends on how they are used; for example, the use of fresh eggs in a product such as a cake that will receive an adequate time–temperature combination during cooking will not be a food safety issue.

Purchasing your ingredients

You can purchase ingredients from suppliers who will deliver to your premises, from fresh-produce markets or from retail outlets such as supermarkets. Regardless of which of these applies, it is your responsibility to do everything within your control to purchase safe ingredients.

Purchasing from a supplier – approved supplier programs

Most fresh-produce merchants at markets will have their own approved supplier programs in place, which are termed vendor assurance schemes. Retail outlets such as supermarkets will also

Box 48 – Illness caused by food containing raw or undercooked eggs

Foodborne disease outbreaks have been frequently associated with the use of raw or undercooked eggs. Foods that have been implicated in outbreaks include aioli dressing, chocolate mousse and uncooked cheesecake. Egg products (whole eggs, whites or yolks that have been separated from the shell) sold in Australia must receive a pasteurisation treatment that will kill pathogenic bacteria of concern. This is a requirement of the Code (Standard 1.6.2).

It is highly inadvisable for fresh, non-pasteurised eggs to be used in products, unless they are heat processed using a time–temperature combination adequate to kill bacterial cells (see under 'Pasteurisation' in Chapter 6).

In some Australian states and territories, specific food safety regulations have been implemented for businesses using fresh eggs in products that will not receive a sufficient heat process. To determine if these apply to you, check with your local food authority.

have their own approved supplier programs in place. Therefore the focus of this chapter is ingredients you source directly from individual suppliers who will generally deliver these to your premises (either themselves or via a third-party).

Your decision to choose one supplier over another should be based not only on factors such as price, reliability and product quality, but also the safety of the items they supply. Because practices used to protect a food's safety will often also maintain its quality, a safe food supplier will generally also supply foods of high quality. For example, storing perishable food at 5°C or below will be beneficial for both quality and safety.

Setting up an approved supplier program will help you to monitor the safety of ingredients you purchase. The level of complexity required for your program depends on the type of ingredients you use in your business, the size of your business and the types of products you make. Businesses using potentially hazardous foods, particularly those that will not undergo a pathogen-control step during or after processing, will need to have more detailed programs. Pathogen-control steps are commonly called 'kill-steps'; see Box 49 for further information.

An effective approved supplier program contains two key components: a method for checking the adequacy of your supplier's food safety controls and the establishment of an approved suppliers list.

Let's look at the different ways you can check your supplier's food safety controls, starting from the simplest through to more complex approaches:

- Ask for copies of your supplier's documentation for any external accreditation of their HACCP plan (Box 50), Food Safety Program or other food safety management systems (such as an allergen-management policy). You should check dates to make sure that they are current.
- Ask to see reports from any independent audits that have been done on the supplier's food safety management system. Results from audits must be acceptable, and they should include checks on the relevant requirements of the Code.
- Ask to see reports from independent testing laboratories that have assessed the supplier's products (called Certificates of Analysis).
- Keep copies of any certificates or documents in your supplier files and request updated versions, either when they expire or at regular intervals.
- You can also send your suppliers a questionnaire asking them how they process the product and the safety procedures they have in place. You can use the information provided in this book to judge the adequacy of the answers. The Australian Food and Grocery Council has developed a standardised supplier questionnaire called the Product Information Form (PIF). Many suppliers will already be familiar with the PIF, and will have all the required information readily at hand. For details about where to find the PIF, refer to page 277.

Once you are satisfied with the results of the checks that you have performed on your suppliers, you can add them to your approved suppliers list. An approved suppliers list is simply a list of all the suppliers that have been through your approval process, alongside all records and documents relevant to your dealings with them. Once you have set up your approved supplier list, you should avoid purchasing materials from new suppliers until they too have passed your approval process.

Regulatory requirements and product specifications

Now that you have a list of approved suppliers, you need to establish product specifications for the ingredients you purchase. Product specifications are a set of requirements or conditions that need to be met to ensure that food is safe to be used and also cover the user's quality requirements.

By law, all food businesses must follow the requirements of the Code in addition to any relevant food regulations specified by state and territory authorities. You may choose to list some of these in your product specifications, particularly those that concern the safety of the ingredients you use. For example, the delivery temperature of potentially hazardous food should be 5°C or below.

Box 49 – Pathogen-control or kill-step

Potentially hazardous foods must be stored under temperature controlled conditions (e.g. 5°C or below) to reduce pathogen growth and/or toxin formation.

Temperature control is particularly important for any potentially hazardous foods used as an ingredient in products that do not undergo a pathogen-control step before consumption. A pathogen-control step is a process used to reduce any pathogens that may be present to safe levels and almost always involves heating. A classical example of this is pasteurisation of milk by heating it using a specific time–temperature combination (i.e. 72°C for 15 seconds, or equivalent combination).

Potentially hazardous foods that may not normally have a pathogen-control step applied to the final product include ready-to-eat chilled desserts, soft cheeses, pre-prepared salads and sliced fruits.

Regardless of whether your product specifications list individual requirements or not, it is usual to include a statement such as 'Must comply with all requirements of the Australian New Zealand Food Standards Code' in every specification.

Examples of the types of things on which it is advisable to state specifications are:

- the delivery temperature of potentially hazardous foods
- the amount of shelf-life you want from the ingredient, so that it does not expire before you are able to use it
- ingredient labelling for food allergen identification
- the type of packaging (if this is important)
- the size or weight of the individual units
- any important properties, such as pH or water activity
- the condition of the product and the package.

Once you have determined which specifications are relevant to the ingredients you are receiving, you should (if possible) establish a signed agreement between your business and your supplier, detailing your required product specifications. Having an agreement means that both parties know where they stand if there is a need to question if the specifications listed have been met. All specifications that are legislated in the Code or applicable state or territory regulations are non-negotiable.

Box 50 – Hazard Analysis Critical Control Points (HACCP)

Many people would have heard of, or seen, the acronym HACCP. This is a food safety management system used extensively by the food industry worldwide. Food Safety Programs, discussed in Chapter 1, are based on HACCP principles, but are usually not as comprehensive as HACCP plans.

Setting up a HACCP plan involves systematically studying all steps in the manufacture of individual products and determining if any significant food safety hazards are associated with them. Each significant hazard identified then has controls specified for it. Those for hazards that are not controlled by support programs are called critical control points (CCPs). Support programs include cleaning and sanitation, and preventative maintenance programs.

Regular monitoring and recording of the CCPs is essential to make sure they are effective and are managed correctly. If something goes wrong, corrective actions need to be implemented and recorded. Verification and validation of the HACCP plan are also important. Verification involves ensuring that you are doing what your HACCP plan says you are doing. A common method is an audit, but it can also involve simple things like checking that record forms are correctly filled in and appropriate corrective actions have been implemented. Validation is the process of confirming that the control you have put into place is correct. Evidence for this can come from reference books, guidelines specific for your product types, results of laboratory testing, or the control is a requirement of the Code or state or territory legislation.

Even very small businesses can establish a HACCP plan. This book may assist you to establish your own HACCP plan. There are also consultants and tools that can help you through the process; see page 277 for details.

Any specifications that are relevant to the safety of potentially hazardous foods that will not undergo a pathogen-control step should be checked every time a delivery of items in this category is received. Delivery temperature, use-by date, and condition of packaging should all meet the specified requirements. Details of these checks should always be recorded and this can be done on the delivery invoice or on a goods receipt form you design yourself.

If a supplier consistently delivers items that do not meet your specifications for food safety, you should, if possible, stop using the supplier. If they are the only supplier you can use, and they do not have a documented food safety system, encourage them to put such a system into place. Ultimately, it is your responsibility to make sure your products are safe for consumption. If you choose to stop doing business with any suppliers, they should be removed from your approved suppliers list and a record kept of the reason why.

To illustrate how product specifications may differ for food types, two examples are provided in Box 51. These are quite detailed to show you the sorts of information you may include. Your specifications may not need to be as detailed, but should at least include all relevant safety requirements. The requirements in your product specifications can be checked against the information provided by the supplier in a PIF, if used.

*

Box 51 – Example product specifications

SMITH & SONS

PRODUCT SPECIFICATION: BACON RASHERS

VERSION NO: 2

EFFECTIVE FROM: 18 June 2009

REPLACES: Version 1, 10 January 2006

REASON FOR CHANGE: Addition of allergen labelling

DESCRIPTION: Triple-smoked short cut bacon rashers, which must comply with all requirements of the Australian New Zealand Food Standards Code

SIZE/VOLUME/COUNT/WEIGHT: 10 × 1 kg packs per carton

PACKAGING: Vacuum packaged in a 'food grade', fit-for-purpose pack. Packed in a cardboard outer pack sealed with glue or tape (NOT staples)

ARRIVAL TEMPERATURE: 5°C or below

LABELLING: Name of product, best-before date, ingredients, usage directions, batch or lot number, country of origin

SHELF-LIFE: Should have at least 2 weeks shelf-life on delivery

COMMON ALLERGENS PRESENT*: Must be declared on label

QUALITY REQUIREMENTS: No evidence of rancidity, slices to be 2 mm thick (between 1.75 to 2.15 mm) and 145 mm long (between 135 to 155 mm)

OTHER REQUIREMENTS: None

APPROVED BY:

Name Position Date

*As per the Food Standards Code, Standard 2.2.1

SMITH & SONS

PRODUCT SPECIFICATION: SHELLED PEANUTS

VERSION NO: 1

EFFECTIVE FROM: 25 May 2009

REPLACES: Not applicable

REASON FOR CHANGE: Not applicable

DESCRIPTION: Raw shelled and hulled peanuts, which must comply with all requirements of the Australian New Zealand Food Standards Code

SIZE/VOLUME/COUNT/WEIGHT: 20 kg bags

PACKAGING: Heat-sealed new polyethylene bags

ARRIVAL TEMPERATURE: To be kept cool

LABELLING: Name of product, best-before date, batch or lot number, country of origin

SHELF-LIFE: At least 9 months shelf-life on delivery

COMMON ALLERGENS PRESENT*: Must be declared on label

QUALITY REQUIREMENTS: Free from obvious mould contamination, no evidence of rancidity, 90% whole kernels, less than 7% moisture, less than 1% foreign matter, free from insect and pest contamination

OTHER REQUIREMENTS: Levels of aflatoxin, agricultural chemicals and cadmium must be within levels specified in Standards 1.4.1 and 1.4.2 of the Food Standards Code.

APPROVED BY:

____

Name Position Date

*As per FSANZ Food Standards Code, Standard 2.2.1

Steps to follow when receiving deliveries

Now that you have established an approved supplier program and product specifications, you and your staff need to put into practice the appropriate checks each time a delivery arrives.

The Code specifies the following requirements for businesses receiving food deliveries:

- A business must take all practicable measures to ensure it only accepts food that is protected from the likelihood of contamination (e.g. check that the packaging is not damaged).
- A business must be able to provide the name and Australian business address of the vendor, manufacturer, packer or importer for all food located on the premises.

Box 52 – Prescribed names for foods

Safe Food Australia defines a prescribed name as, '… a name that has been legally specified for a food. For example, "milk" is a prescribed name. Most foods will not have a prescribed name and instead will be identified by an appropriate designation or a common name. An appropriate designation specifies what the food is, for example chocolate dairy dessert'.

Packaged food that is sold to food manufacturers (i.e. not retail sale) must be labelled with the name of the food and the supplier. However, as this information is only required to be on the outer packaging and not on the individual packages of food within, when the outer packaging is discarded this information is no longer available. If a raw ingredient used by a food business is recalled, the business operators must be able to quickly determine if they have this product on their premises. That is why keeping track of items purchased from each supplier is important. A simple way to do this is to file supplier invoices in a systematic way.

- A business must be able to provide a list of prescribed names or an appropriate description of all food items located on the premises (see Box 52).
- A business must, when receiving deliveries of potentially hazardous foods, take all practicable measures to ensure that it only accepts items that are:
  - 5°C or below; or
  - 60°C or above

  unless the business transporting the food is able to prove that transport between 5 and 60°C for the time taken to transport the food, will not adversely affect the microbial safety of the food.
- A business must, when receiving deliveries of potentially hazardous foods, take all practicable measures to ensure food that is intended to be received frozen is frozen when it is accepted (i.e. hard or solid throughout).

Source: Standard 3.2.2 Food Safety Practices and General Requirements

Goods receipt form

You can fill in a goods receipt form each time you receive a delivery of ingredients or you can write the required information onto the delivery invoice (Box 53). Either way, you will have a record of the condition of the goods when you received them.

Box 53 – Example goods receipt form

SMITH & SONS

GOODS RECEIPT FORM FOR POTENTIALLY HAZARDOUS FOODS

Potentially hazardous food received: raw beef and lamb; ham; milk; fresh pasta

Criteria for acceptance/rejection of delivery:

Temperature upon receipt: Refrigerated items must be 5°C or below – if more than 8°C reject; if 5–8°C, call supervisor for corrective action instructions

Use-by time remaining upon receipt: Must have at least 5 days remaining to use-by date (including the day received), if past use-by date reject; if less than 5 days remaining, call supervisor for corrective action instructions

Packaging upon receipt: Outer packaging should have no signs of damage or product leaks; if evident, open and inspect inner packaging – if inner packaging damaged or leaking, reject

Truck condition – Must be clean and have no signs of pest infestation – if not clean, call supervisor for corrective action instructions; if pests evident, reject

Date received (2009)	Supplier	Product code and type	Temp. °C	Use-by date	Packaging condition	Truck condition	Accept	Rejection reason	Initial
10 Sep	*Johnsons*	250808 *diced lamb*	*4*	*15 Sep*	*Pass*	*Pass*	*YES*		*BB*
15 Sep	*Small-goods Co.*	*5248 sliced ham*	*11*	*24 Sep*	*Pass*	*Pass*	*NO*	*Temp. abuse, SC notified by phone*	*BB*

Having a goods receipt form will benefit you in several ways:

- It will reduce the risk of overlooking something important by providing a list of product specifications or other requirements the person receiving the delivery needs to check.
- It allows you to keep a record of the condition of the product upon receipt, such as its temperature, should a problem later arise.
- If the delivered items do not meet specifications, you may choose to reject them. Recording this information on the goods receipt form allows you to build-up a 'history' on your suppliers, helping you identify any who you wish to discontinue using.
- If necessary, you or the authorities can more effectively trace the cause of any food safety issues linked to your product ingredients through records of batch codes and/or date marking.

Overview of what to check for

Checking the delivery can commence as soon as the delivery truck arrives. Here are some tips on what to look for:

- Does the truck appear clean and well maintained? Take note of the following for refrigerated trucks:
  - excessive ice build-up on the inner walls
  - fridge unit making unhealthy 'rattling' sounds
  - damaged seals around the door.
- Are there any off odours or chemical smells coming from the truck or from the delivered items?
- Are outer boxes or packaging torn, damaged, wet or dirty?
- Do cartons or packages of chilled foods feel cold to the touch?
- Are there any signs frozen foods have thawed; for example, wet or water damaged cardboard boxes?
- Is there a use-by date printed on outer packaging (when this is appropriate to the food type), and is this sufficient for your requirements? Refer to Box 54 for further information.
- Is other required information clearly readable on the outer packaging, including the name of the ingredient and the supplier?

If you purchase any packaging materials that will come into direct contact with food, such as plastic tubs or foil trays, you should also check that their outer packaging is clean and undamaged.

Other checks can generally only be performed once the outer packaging is opened:

- Are there any signs of pest damage or insect infestation?
- Are there any leaking, torn or swollen packages?
- Are there any swollen, rusty or dented cans?
- Are there any visibly spoiled food items?
- Are there any signs frozen foods have thawed and then been refrozen? Examples of this include ice crystals on the packet or surface of food, food no longer in its original shape or food clumped together.

Special care should be taken when receiving deliveries of potentially hazardous foods. Staff members receiving these deliveries should know how to check the delivery temperature and the importance of quickly transferring the items to the appropriate storage area (e.g. area designated for raw meat in a cold room). See the next segment, 'Checking the temperature of potentially hazardous foods', for further details.

Box 54 – Specifying shelf-life requirements for delivered items

If you are receiving raw ingredients that have a use-by or best-before date, you do not want this date to pass before you have an opportunity to use the item. This is critical for use-by dates because these are specified for food safety reasons (Chapter 7). You should immediately reject any items that are delivered after their use-by date has passed.

Your product specifications should state how many days, weeks or months you require to have remaining in the use-by or best-before period at the time of delivery. These shelf-life requirements should also be included on goods receipt forms so the dates are checked and recorded. This is important for any ingredients that have a use-by date, because you want to be sure that you have sufficient time to use them before they must be discarded.

Foods that have a shelf-life of 2 years or longer, including most canned foods and some dry products, are not required to have a date marked on them. You may wish to mark these with a date yourself to assist you to track which items should be used first.

Checking the temperature of potentially hazardous foods

As specified in the Code (Standard 3.2.2), it is a legal requirement when receiving deliveries of potentially hazardous foods to check they are not delivered at a temperature within the temperature danger zone (between 5 and 60°C; Box 55). As it is highly unlikely that a food manufacturing business would require delivery of any pre-heated ingredients, the focus of discussion here is on potentially hazardous foods that need to be transported at 5°C or below.

If potentially hazardous food is above 5°C when it is delivered to you, it may need to be rejected, unless you are satisfied the food will still be safe; for example:

- use of the 2 hour/4 hour guide temperature control system (Box 56)
- use of another temperature control system documented as part of a food safety management system, with results for laboratory tests performed available as proof that the method is safe
- following a procedure in a relevant industry guideline.

If your temperature measurement shows that goods are above 5°C upon receipt, you have the right to ask the delivery driver to show you the data logs for the temperature monitoring inside the truck; then you can see for how long the goods have been at an air temperature above 5°C. You should consider rejecting the delivery if the driver can't or won't show you this information.

This consequence should have already been discussed with your suppliers when you communicated to them your product specification requirement for potentially hazardous food to be delivered at 5°C or below.

It is your responsibility to accept only deliveries of ingredients that are safe to use in your products. The Code, however, does use the term 'practicable measures' when describing what the requirements are for checking the delivery temperature of potentially hazardous foods. This means that is not necessary for you to individually measure the temperature of every potentially

Box 55 – Why is between 5 and 60°C the temperature danger zone?

Pathogens cannot grow at temperatures above 60°C. Below 5°C, some pathogens can grow but only very slowly. Potentially hazardous foods should not be stored between 5 and 60°C for long enough to allow any pathogens that may be present to grow to unacceptable levels (i.e. likely to cause illness).

Storage at 5°C or below is most appropriate for foods that are intended to be served cold or at room temperature, but that are not going to be eaten straight after preparation. Food served hot, such as meals in a food service situation, must be kept at 60°C or above until it is either eaten or discarded.

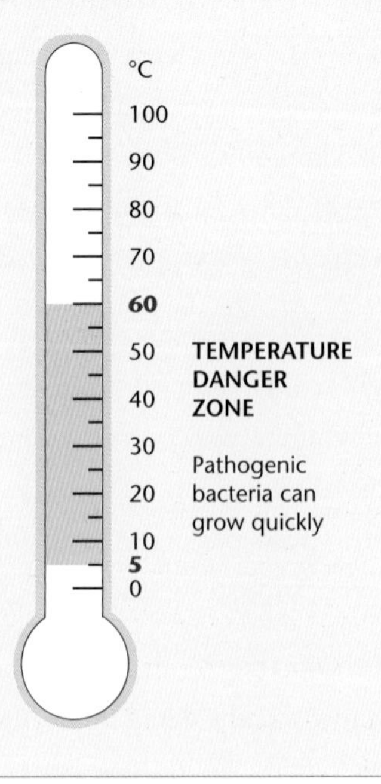

Box 56 – The 2 hour/4 hour guide for temperature control

It may not be possible for food businesses to always store, transport or prepare potentially hazardous foods at 5°C or below, or at 60°C or above. You are permitted to have these foods at other temperatures, as long as you can show that this will not make them unsafe. The key is to limit the overall amount of time these foods are in the temperature danger zone.

The 2 hour/4 hour guide has been developed to provide practical limitations on the time potentially hazardous foods can be kept in the temperature danger zone:

The 2 hour/4 hour guide

Total time between 5°C and 60°C	You must:
Less than 2 hours	Refrigerate at 5°C or below or use immediately
Between 2 and 4 hours	Use immediately, DO NOT refrigerate for later use
More than 4 hours	DO NOT USE (discard)

These time limitations are for the total length of time food is above 5°C or below 60°C. For example, if it takes you 1 hour to transport (un-refrigerated) food to your premises and then 1.5 hours to prepare it – it has been at room temperature for longer than 2 hours in total and so must be used within 4 hours, and not chilled for later use.

More information on how to apply the 2 hour/4 hour guide can be found in the FSANZ publication, Food Safety: Temperature control of potentially hazardous foods (refer to page 272 for details).

hazardous food item that is delivered to you; it allows for some flexibility. For example, when receiving the first few deliveries from a new supplier you may wish to perform temperature checks on every delivery. Once you have established confidence that this particular supplier, or their delivery service, consistently delivers items at 5°C or below, you may reduce the frequency of checks. In addition, if receiving several cartons in one delivery, you may only need to check the temperature of items in the carton most likely to be temperature abused (i.e. usually a carton near the door of the delivery vehicle).

Probe thermometers

If you store, transport, prepare, cook or sell potentially hazardous foods, it is a requirement of the Code that you have a thermometer with a probe that can be inserted into food. This must be accurate to within 1°C (plus or minus); see Chapter 3 for further information. When receiving deliveries of potentially hazardous food, you can use this thermometer to check the delivery temperature of items without opening up individual food packets or containers. This means that the food does not need to be exposed or touched, avoiding the risk of contamination.

The basic steps are:

- Preferably, measure the temperature in a shaded area.
- Open the outer carton and then place the length of the thermometer between two packets (to the depth specified in the operating instructions).
- Press the two packets together so the thermometer probe is in contact with the surface of the packets.
- Wait for the time specified in the instructions for the thermometer to have enough time to measure the temperature accurately (i.e. to equilibrate).
- Read the temperature and record the result on your goods receipt form (or delivery invoice).
- Clean and sanitise the thermometer probe before using it to measure the temperature of any unpackaged foods; see Chapter 6 (Box 61) for the method.

Infra-red thermometers

Although infra-red (IR) thermometers allow users to measure the temperature of the outer surface of packets of food quickly, by pointing a laser beam at the item, they are not always consistently accurate enough to be used for measuring the temperature of potentially hazardous foods. This is because variables such as the amount of light present (i.e. the difference between a sunny or overcast day) the reflective properties of the packaging used or the distance the gun is held from the food, can cause large differences between 'real' and 'measured' temperatures.

Time–temperature indicators

Your supplier may use time–temperature indicators to monitor the temperature of products they deliver. These are disposable strips that are normally attached to the outside of cartons. However, you should still use your own thermometer to measure the temperature of items inside cartons. Cardboard boxes, or other outer packaging, can act as insulation because they can 'trap' warm air inside. The temperature of the outside of the carton does not necessarily match the temperature inside the carton.

Data loggers

Temperature data loggers are very useful for following what happens to temperature over time and, because they are fully automatic, they do not require a person to be present to take readings and will operate for 24 hours a day, 7 days a week. Data loggers can give a very accurate picture of the temperature changes within a product. They are usually battery operated and portable. The time interval between temperature readings can be set from between a few seconds to every few days, depending on the situation. Some data loggers have built-in alarms to notify you that there is a problem; some are designed to transmit temperatures back to a receiver so that temperatures can be checked at a remote location as they are being measured.

Checking food for adequate protection against contamination

The Code specifies that a food business '... must take all practicable measures to ensure it only accepts food that is protected from the likelihood of contamination'. This relates to contaminants that can potentially cause food safety issues: pathogenic microorganisms and chemical or physical contaminants.

Inadequate protection from contamination includes:

- unwrapped or unpackaged items
- holes in packaging due to poor handling or insect/pest infestation
- leakage of food out of packaging.

Here, use of the term 'practicable measures' means that businesses are allowed some flexibility. For example, once you are confident a supplier is consistently delivering goods that are adequately protected from contamination, you may chose to reduce the number of checks you perform on their goods.

Traceability

It is a requirement of the Code (Standard 3.2.2) that you know the name of all ingredients used by your business, as well as where these came from. This is required because in the event of a food safety recall of any ingredients, you should be able to correctly identify these and stop using them in your products. Additionally, you must be able to identify any batches of your products that contain the recalled ingredient.

Alternatively, if your own products cause foodborne illness as a result of the use of unsafe ingredients, information you provide to authorities about the source of these ingredients may allow them to act to prevent further cases of illness occurring.

If you discard outer product packaging that states the name of the product and its supplier, you need to have an alternative way of keeping track of these details. For example, filing delivery invoices in a central location.

Use of your own vehicles when purchasing ingredients

In addition to delivered ingredients, you may also need to go to markets or retail outlets to purchase your stock. In this case, you are responsible for temperature control of potentially hazardous foods that you purchase, and for protecting this food from possible contamination (Box 57).

You will need to consider how you will control the temperature of the food during transport and how you will measure that you are achieving this adequately. If you need to transport only relatively small amounts of food over short distances then you may find using insulated foam boxes or cooler bags will suffice. Cool packs, which are also known as freezer bricks, can help to keep products chilled.

Box 57 – Requirements for businesses transporting food

The Code outlines specific requirements for the transport of food by a food business:

- Food must be protected from possible contamination.
- Potentially hazardous food must be transported under temperature control, that is, 5°C or below, 60°C or above, or another temperature for a specified time proven not to adversely affect the microbial safety of the food.
- Potentially hazardous food that is intended to be transported frozen must remain frozen during transportation.
- Vehicles used to transport food must be designed and constructed to protect food if there is a likelihood of food being contaminated during transport.
- Parts of vehicles used to transport food must be designed and constructed so that they are able to be effectively cleaned.
- Food contact surfaces in parts of vehicles used to transport food must be designed and constructed to be effectively cleaned and, if necessary, sanitised.

Source: Standard 3.2.2 Food Safety Practices and General Requirements

If you routinely need to transport larger volumes of chilled foods, you can have eutectic refrigeration systems fitted to small vans. These systems can be run off 12–24 volt batteries. If you also need to make regular temperature controlled deliveries of your products to your customers, you may find it worthwhile to purchase a small refrigerated truck or van. It is then your responsibility to routinely monitor the temperature inside these vehicles, just as you do for cold rooms or fridges inside your premises.

If picking up or shopping for multiple items at different locations, plan your route so that any items that need to be held at 5°C or below are picked up last. This is especially important in standard vehicles not fitted with refrigeration units.

If you are unable to maintain the temperature of potentially hazardous chilled foods at 5°C or below during transport, you will need to abide by the 2 hour/4 hour guide.

You must also reduce the risk of foods you transport becoming contaminated with microorganisms, chemical or physical contaminants:

- You should have a separate vehicle dedicated for food transport. However, this may not always be possible for small businesses starting out and they may need to use their private or family vehicle.
- If using a private vehicle that is also used for other activities (such as transport of pets or gardening supplies), it needs to be cleaned appropriately before using it to transport food for your business. Food should always be transported in dedicated sealed containers, which are removed from the car if it is used for other activities.
- Domestic pets or farm animals should never be transported inside the same vehicle at the same time as food for your business.
- If transporting items that are not pre-packaged (e.g. fruits and vegetables), you should make sure they are adequately contained during transport. Storage containers should be regularly cleaned and sanitised.
- If you need to transport cleaning products or pesticides at the same time as food, these must be well contained and separated from food items.
- The cargo area of the vehicle should be included in your cleaning and sanitising program.

Storing your ingredients safely

Now that you have safely received or shopped for your ingredients, you need to store them safely. The Code contains specifications for the safe storage of foods (Box 58).

The most important step for the safe storage of food is to store it at the correct temperature for the appropriate amount of time, so any pathogenic microorganisms present are unable to grow to unacceptable levels.

Other recommended practices:

- Store raw meat, poultry and seafood separately from other foods, particularly those that will not be cooked before eating, to reduce the risk of cross contamination.
- Prevent cross contamination by transferring foods into alternative storage containers if the original packaging will not adequately reduce the risk. For example, raw meat that may leak juices onto other foods or surfaces should not be stored in thin plastic bags, which could easily leak.

- Adequately cover or contain foods to prevent the risk of physical or chemical contamination.
- Record information such as the batch number, use-by or best-before dates and storage instructions, if the food is removed from its original packaging.
- Wash hands after handling packaging for raw foods, particularly meat and poultry, as it may be contaminated by pathogenic bacteria.
- Establish a first-in-first-out stock rotation system so you and your staff know in which order to use multiple stocks of the same ingredient or when to discard potentially hazardous foods (see 'Stock Rotation' later in this chapter).

Box 58 – Requirements for businesses storing food

The Code outlines specific requirements for storage of food by a food business:

- Food must be protected from possible contamination.
- Environmental conditions under which the food is stored must not adversely affect the safety and suitability of the food. These include temperature, humidity, lighting conditions and atmosphere.
- Potentially hazardous food must be stored under temperature control.
- Potentially hazardous food that is intended to be stored frozen must remain frozen during storage.

Source: Standard 3.2.2 Food Safety Practices and General Requirements

Recommended storage practices for different food categories

Different categories of food need to be stored in different sections of storage areas. To avoid confusion, labels or signs should clearly show where foods in different categories should be stored. If you can't adequately separate your ingredients in your current storage areas then it is better to scale up your storage capacity.

Potentially hazardous food

All potentially hazardous foods that require refrigerated storage to prevent the growth of pathogenic microorganisms must be stored at 5°C or below as soon as they have been checked following delivery. This is a requirement of the Code.

Perishable foods in this category may include:

- raw meat, poultry and seafood
- processed foods with a pH above 4.6 and water activity above 0.85
- fruits and vegetables that have been pre-prepared (e.g. salad mixes, cut fruit)
- semi-cooked or raw dough and pastries
- soft cheeses, such as camembert or ricotta
- cured meats, such as ham or corned beef.

All raw foods should be well contained and separated from any ready-to-eat foods. Designated storage areas should be established in fridges or cold rooms to aid physical separation of raw foods, such as meat, from prepared foods, such as salads. If you have more than one fridge, ideally dedicate one to raw and one to ready-to-eat foods. If you have only one fridge or a small cold room, store raw food below ready-to-eat foods so that any accidental drip from raw foods does not contaminate ready-to-eat foods.

Although 5°C or below is the temperature specified in the Code for storage of potentially hazardous foods, maintaining temperatures as close to 0°C as possible is desirable. Avoid overcrowding in your fridges or cold rooms, as this prevents the cold air from circulating effectively.

You must establish a stock rotation system for potentially hazardous foods to prevent them from accidentally being used beyond their use-by date or the date set by your own date marking system (see 'Stock rotation' later in this chapter).

Potentially hazardous foods can also be frozen if they will not be used within a few days of receipt or within their use-by date. This is useful for foods that you receive in bulk, but not all of which is required for immediate use (e.g. minced meat). Because freezing does not kill pathogens – it only stops them from growing – it is important that any use-by time remaining on the food at the time of freezing is still followed after thawing. For example, an item with 2 days left on its shelf-life at the time of freezing should be used within 2 days of defrosting. Due to damage to their quality (not safety), some items such as eggs in shells, canned food, cream sauces, mayonnaises and salad items are not suitable to freeze. If you are unsure, you should ask your supplier if frozen storage is suitable for particular items.

Frozen foods

All frozen foods should be kept in their frozen state unless they are going to be used within a short period of time (e.g. 3 days or less). Foods that have been removed from their original packaging should be labelled to indicate their 'age' so that older stock can be distinguished from fresher items.

Freezers should not be overfilled because this prevents cold air from circulating around the food. Freezers that don't have an automatic defrost cycle should be regularly defrosted because built-up ice makes it hard for the motor to maintain the correct temperature. Freezers should be included in your cleaning and sanitation program.

A thermometer or other temperature measuring device should be used to monitor the air temperature routinely. For optimum shelf-life of frozen foods, this should be –18°C or less.

Food allergens

If you manufacture products that are allergen-free and products that are allergen-containing, you must take extra precautions when storing your ingredients. Ingredients containing food allergens must be stored in a way that prevents cross-contact contamination with ingredients that do not contain allergens.

For example, if all your products contain peanuts, then it is not necessary for you to store your peanuts any differently from your other ingredients. However, if some of your products contain peanuts as an ingredient and other products do not, you will need to store the peanuts in dedicated, clearly labelled and sealed storage containers. The information on the label should include the name of the ingredient, an allergen presence warning and the allergen type.

The risk of allergens contaminating other foods owing to a spill or leak can be reduced by storing allergen-containing foods below other ingredients or placing within a second, unbreakable well-sealed container. If a spill or leak of an allergen occurs, you must immediately clean the area thoroughly, including a wide zone around the site of contamination.

Different types of allergens should also be kept separated from each other, because individual people will generally only be allergic to one or two types. Because even tiny traces of some

allergens can cause serious illness if eaten, it is important to avoid cross contact between allergens and allergen-free foods. For example, a scoop used to ladle peanuts out of a storage bin must not be used with any non-peanut containing food until it has been thoroughly cleaned.

Fresh fruit and vegetables

Fresh fruit and vegetables may be contaminated with dirt and/or insects. Before storing, check for signs of damage, rot or insect infestation. Discard any items that are in poor condition.

Items not normally kept chilled, such as onions, should be stored away from other food in a cool, dry and well-ventilated area. Keeping off floor level and placing in storage bins will help reduce the risk of pest and insect infestation.

Refrigerated fruits and vegetables should be kept in bags or containers, to minimise the chance of cross contamination between other foods and surfaces.

Dry and shelf-stable foods

Dry and shelf-stable foods include flour, rice, sugar, unopened cans, and non-perishable foods in bottles, jars or cartons. These items should be stored off the floor on racks or shelves in a cool, dry and well-ventilated area. When possible, glass containers should be avoided, but, if they are used, they should be placed in a secondary container on a lower shelf to minimise the risk of breakage and contamination of ingredients.

You must check for, and follow, any specific storage instructions on the packaging or label. Once packets of dry goods are opened, it is best to transfer the contents into a sealable 'food grade' container, to reduce the risk of insect or pest infestation.

Any leftover contents of opened cans should be transferred into a sealed 'food grade' storage container and stored in the fridge. High-acid foods (below pH 4.6) may be kept refrigerated for several weeks. However, leftover low-acid foods (above pH 4.6) are generally potentially hazardous and should be stored for a more limited time, say 3 days (at 5°C or below).

Key information from labels, such as batch numbers and allergen warnings, should be transferred onto the storage containers.

Labels on shelf-stable foods contained in bottles, jars or cartons should be checked to see if they contain instructions to store chilled after opening. If no instructions are provided, it is best to err on the side of caution and store the product in the fridge after opening. However, high-acid foods, such as tomato sauce, may be safely kept at room temperature after opening.

The food storage area should be regularly swept out and cleaned, and any spilt food should be cleaned up immediately.

Food storage containers and covers

Any material that comes into direct contact with food must be 'food grade'. Plastic food packaging materials available for use in Australia must meet the requirements of the Australian Standard for Plastics Materials for Food Contact Use (page 273). If the manufacturer's

instructions for use are followed, contamination of food with unacceptable levels of chemicals should not occur.

When not in use, food storage containers and covers must be protected from contamination by microorganisms, chemicals or physical contaminants. They should be stored in a clean dry area, free from pests and insects and separated from food preparation areas.

Re-using disposable packaging or using packaging for anything other than the original purpose is not recommended. Plastic wrap, cardboard containers or other packaging that cannot easily be cleaned should never be re-used. Packaging such as rigid plastic containers can be re-used if thoroughly cleaned and sanitised in-between uses. However, if these become scratched, cracked or otherwise damaged, they should be replaced as they may no longer seal or be sanitised effectively.

The same container type that is used for sale and distribution of your final products should not be used to store raw ingredients or 'in-process' products. This avoids potential confusion and the possibility of unprocessed product accidentally being distributed.

Re-packing and portioning bulk ingredients

Bulk packs of foods, including meat or fish, can be divided into smaller portion sizes as required in your recipes. This is useful if you will be freezing these items to be used later, as smaller packs will thaw faster. It is also useful for foods such as pasteurised egg products, which may not be available in small packs. Splitting up items like this into smaller portions reduces the need to pull a bulk pack in and out of the fridge or freezer several times, and reduces the chances of cross contamination caused by opening and closing packaging many times.

You should take precautions when handling ingredients during this process; for example:

- use clean utensils and packaging material or containers
- reduce the total amount of time any potentially hazardous foods are at room temperature (i.e. follow the 2 hour/4 hour guide)
- label portions appropriately, including use-by dates and allergen information if required
- make sure your stock rotation system is applied.

Stock rotation

You must develop a system that allows you and your staff to know which ingredients should be used first, and when any potentially hazardous foods should be discarded. There is no need to make this overly complicated: it should be simple and practical. The key is to be certain that all your staff understand how the system works.

This also applies to your own internal use-by date system for ingredients that you prepare yourself that will not be used immediately in your final product; see Box 59 for an example.

The system must be strictly followed for any potentially hazardous foods that will not be cooked thoroughly before being eaten. Some types of pathogenic bacteria are able to grow slowly at

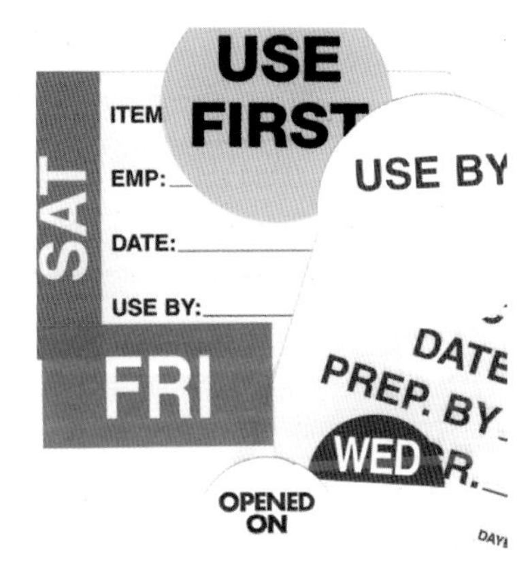

refrigeration temperatures, so there is still a need to limit the amount of time certain foods are kept in the fridge.

Use-by dates provided on packaged foods are present for food safety purposes: they must be followed (see Chapter 7 for further information). It is good practice at the end of each day to check the fridge or cold room for any items that have reached their use-by date (including your own internal dates), and discard these immediately so they are not used accidentally. If you find that you are often discarding raw ingredients that have reached their use-by dates, you should reduce your order volumes.

You can achieve adequate stock rotation by using a combination of date or day of receipt coding and physical placement (e.g. placing older stock in front of newer stock). A useful method for high turnover chilled ingredients is to use a different coloured sticker for each day of the week to show when items were received.

Separation of rejected or discarded foods

When foods go past their use-by date, or are recalled by their supplier, they must be isolated and, if not immediately discarded, clearly labelled so they do not accidentally find their way into your products. The label should state the reason why the item is not to be used (see photo below). Using a label that simply says 'Do not use' or 'Hold' may be mistaken for food that is being reserved for later use.

Visibly spoiled items should be well contained to prevent cross contamination; for example, a swollen pack may burst open and spread contaminants in the air, which can transfer to other foods in the area.

Box 59 – Cooked rice not used on the same day it is prepared

Owners of a fruit and vegetable shop make their own range of salads and routinely cook large batches of rice for later use as an ingredient in a rice salad product. Cooked rice is a potentially hazardous food, so safe handling, including temperature control, is important.

Firstly the owners need to make sure the cooking and cooling process is safe (Chapter 6). Secondly, they need to use a system that restricts the length of time the rice is kept refrigerated before use in the salad mix.

This is an example of the type of label that can be used:

This batch of rice was cooked on a Friday and, if not fully used by close of business Sunday, it must be thrown out.

Note – A shelf-life restriction would also need to be placed on the refrigerated storage time of the final rice salad product. This will be discussed in Chapter 7.

A record should be maintained for any ingredient spoilage incidents that do not match normal expectations. For example, you would expect bread to go mouldy if stored at room temperature for over a week in a humid climate, but an off-odour in a carton of shelf-stable chicken stock still within its best-before date is abnormal. Keeping a record will assist you to discuss the issue with your supplier or the manufacturer, if desired. Some manufacturers may want to test spoiled products, so it might be worthwhile contacting them before discarding the item.

Requirements of the Code – food disposal

The Code, Standard 3.2.2, specifies requirements for food that is deemed to be 'for disposal'.

In this category are foods that:

- are subject to recall
- have been returned to you
- are not safe or suitable, or
- are reasonably suspected of not being safe and suitable.

These foods must be held and kept separate from other foods, or food handling areas, until destroyed (i.e. there is no chance of them being eaten), returned to the supplier, processed in a way that ensures its safety and suitability, or determined to be safe and suitable.

These foods must be clearly labelled to identify why they must not be used, such as by using the words 'recalled', 'for return to supplier' or 'unsafe'.

KEY MESSAGES FROM CHAPTER 5

- Approved supplier programs allow for greater control over the safety of raw ingredients purchased.
- Product specifications should include key requirements that need to be met for food to be safe and suitable for use.
- When receiving deliveries, the requirements of the Code must be met; for example, accepting only food protected from contamination.
- The length of time potentially hazardous food is stored between 5 and 60°C (the temperature danger zone) must be kept to a minimum.
- The 2 hr/4 hr guide provides guidance on timeframes food may safely be left in the temperature danger zone.
- An effective system to trace which ingredient batch is used in which final product batch must be used.
- When transporting food, the requirements of the Code must be met; for example, potentially hazardous food must be transported under temperature control.
- When storing food, the requirements of the Code must be met; for example, food must be protected from possible contamination.
- Different food types must be adequately separated during storage; for example, raw meat from salad vegetables.
- A system must be used to make sure that ingredient stocks are rotated appropriately and that use-by dates on ingredients are strictly followed.
- Foods that are deemed unfit for eating (or are pending assessment), must be clearly labelled and kept separate until disposed of. This is a requirement of the Code.

Chapter 6

Controlling food safety hazards – preparing, cooking and cooling safely

So far, we have considered how you and your staff, your premises, your product recipes and the ingredients used can affect the safety of your products. Let's now look at practices during the preparation, cooking and cooling phases that can play a major role in enhancing the safety of food products.

The planning phase – it pays to think before you make

Writing down (documenting) how you prepare your products in the form of a flow chart allows you to think through each step individually and record any specific practices that are necessary to reduce the risk of cross contamination (of pathogens or allergens). On the following page is an example of a flow chart used to describe a food preparation process. For simplicity, every detail has not been shown. In practice, much more specific information should be included, such as the length of time required for thawing of chicken.

Basic steps for safely preparing food

Food preparation consists of all the tasks needed to get your ingredients in the right state and correct combinations before packaging and/or cooking. These tasks include chopping, grating, mixing, portioning, seasoning, and so on.

It is essential that you follow good hygiene practices during all the preparation stages, even if you cook your product post preparation. It is a common misconception that, if products will

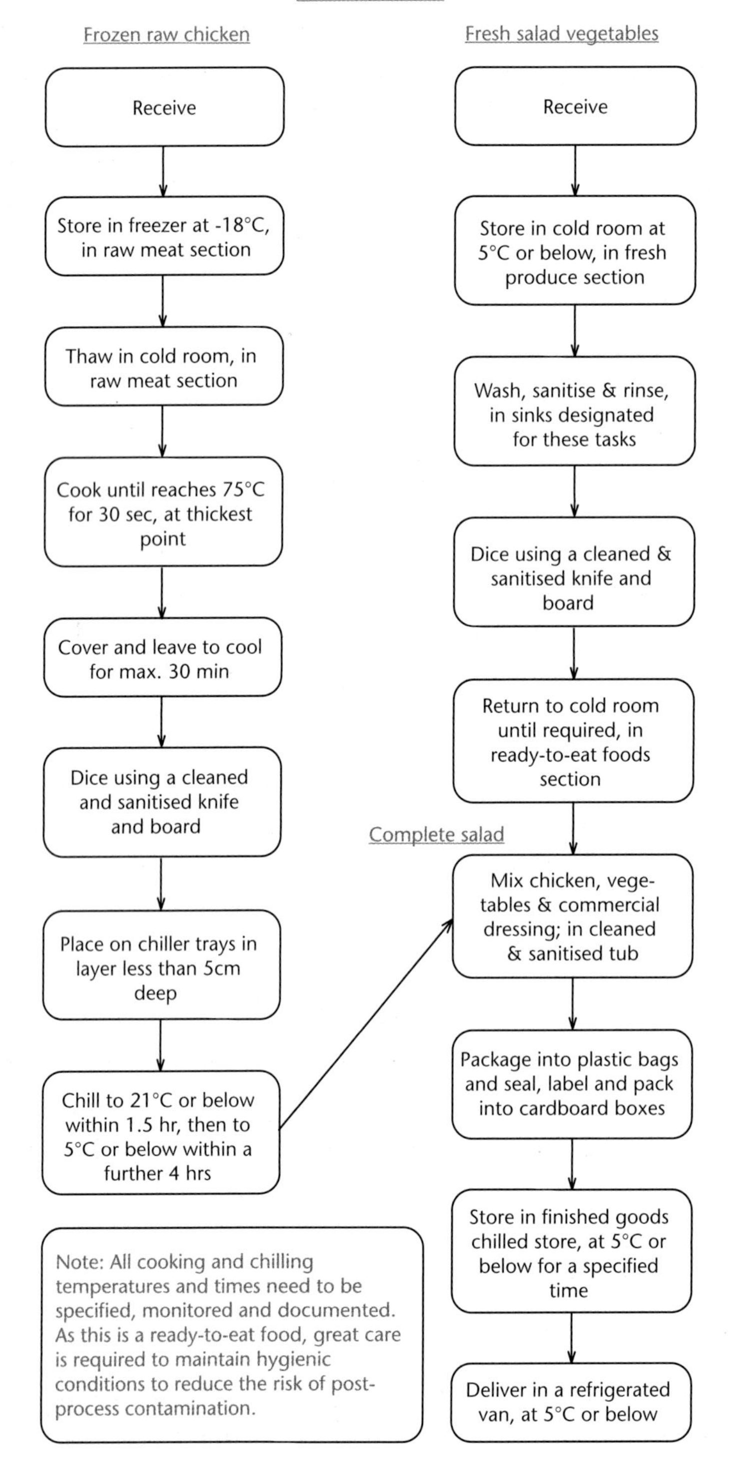
CHICKEN SALAD
Frozen raw chicken
Fresh salad vegetables
Receive
Receive
Store in freezer at -18°C, in raw meat section
Store in cold room at 5°C or below, in fresh produce section
Thaw in cold room, in raw meat section
Wash, sanitise & rinse, in sinks designated for these tasks
Cook until reaches 75°C for 30 sec, at thickest point
Dice using a cleaned & sanitised knife and board
Cover and leave to cool for max. 30 min
Return to cold room until required, in ready-to-eat foods section
Dice using a cleaned and sanitised knife and board
Complete salad
Mix chicken, vegetables & commercial dressing; in cleaned & sanitised tub
Place on chiller trays in layer less than 5cm deep
Package into plastic bags and seal, label and pack into cardboard boxes
Chill to 21°C or below within 1.5 hr, then to 5°C or below within a further 4 hrs
Store in finished goods chilled store, at 5°C or below for a specified time
Note: All cooking and chilling temperatures and times need to be specified, monitored and documented. As this is a ready-to-eat food, great care is required to maintain hygienic conditions to reduce the risk of post-process contamination.
Deliver in a refrigerated van, at 5°C or below

be cooked, little care is required during handling steps pre-cooking. This is not true. Cooking does not always kill all pathogens or destroy all heat resistant toxins, so it is important to keep the levels of contaminants in raw foods to a minimum. Additionally, temperature abuse during preparation may cause pathogen and toxin levels in the food to rise, increasing the risk of the cooking process failing to reduce contamination to acceptable levels. Further, cooking will only affect the microbial hazards; you also need to consider allergens, chemical and physical hazards.

Chilling as you go

Product ingredients that need to be kept refrigerated must not be kept out of the fridge or cold room for long periods during product preparation. These foods include potentially hazardous foods, anything with a use-by date and items labelled 'Store at 5°C or below'(or similar).

Remember, between 5 and 60°C is the temperature danger zone and the 2 hour/4 hour guide should be followed when handling these foods at room temperature. Any potentially hazardous food that is left out of the fridge for more than 4 hours should be discarded, even if it is still to be cooked. If your product will be fully cooked after preparation, then you can start timing a new 4 hour period after the cooking step.

One way to minimise the length of time food is in the temperature danger zone is to split-up bulk tasks into smaller sub-sets. For example, if you were shelling cooked prawns to be used in a seafood salad, you may divide them into batches that you can shell within (say) 30 minutes and then shell one batch at a time while the other remains in the fridge.

Washing and sanitising fruit and vegetables

The aim of washing fresh produce is to remove physical contaminants such as dirt and insects. Fresh produce that will be eaten without any cooking step should also be sanitised. This aims to reduce the level of pathogens that may be present. Examples of produce that should be sanitised are those used to make:

- salads
- dips and sauces that will not be cooked
- garnishes (e.g. fresh herbs or cut fruit).

The basic process is essentially the same as cleaning and sanitising equipment and surfaces, with steps 3 and 4 optional (depending on the product you are making):

1. **Prepare** – remove damaged parts and obvious physical contaminants, separate leaves, and so on.
2. **Wash** – in clean water, replace the water if it becomes so dirty it no longer appears to be washing effectively.
3. **Sanitise** – soak for 5 minutes in 0.01% (100 ppm) chlorine; the same concentration recommended for sanitising food contact surfaces (Chapter 3, Box 29).
4. **Rinse** – remove the sanitiser in clean water.
5. **Separate** – keep covered and separated until use so that re-contamination does not occur.

It is best to wash and/or sanitise fruit and vegetables before cutting or shredding to reduce the spread of dirt or pathogens deep into the flesh of the produce where they are very difficult to remove.

The washing, sanitising and rinsing steps can be performed in the same sink, but it should be a different sink from the one used for cleaning dirty utensils, and so on. If this is not possible, then you may choose to use a large tub instead of a sink. After use, the sink or tub needs to be cleaned and sanitised. Under no circumstances should produce be washed or sanitised in a basin used for washing hands.

It is important to have the temperature of the water used for washing and sanitising about 10°C warmer than the produce. This reduces the amount of water that is absorbed into the produce, possibly carrying pathogenic microorganisms.

The concentration of chlorine can be measured using dip sticks available from laboratory suppliers. It is recommended that a table is used to record the details, so you can check if the sanitising process is being performed correctly by staff.

The table should include the:

- date
- chlorine concentration (actual)
- dipping time
- staff members name or initials.

Alternative chemicals to chlorine are available for sanitising produce. You must only use 'food grade' substances that are listed as permitted for use in the Code. All substances must be used strictly according to the instructions provided with the product. A wetting agent may be added during the sanitising step to aid the effectiveness of sanitisers. A wetting agent permitted for use in the Code is sodium lauryl sulphate.

This information about washing and sanitising fresh produce was adapted from the NSW Food Authority's Vulnerable Persons Food Safety Scheme Manual (page 273).

Thawing frozen foods

The time potentially hazardous foods are in the temperature danger zone should be kept to a minimum. As food will thaw from the outside first, the outer sections of foods thawed at room temperature can be in the temperature danger zone, while the centre is still frozen solid. This is highly risky for ready-to-eat, potentially hazardous food that will not be cooked or otherwise processed before being eaten. These foods should generally be thawed in a fridge or cold room at 5°C or below. However, if thawed at room temperature, the 2 hour/4 hour guide should be followed.

Frozen raw red meat and poultry can be safely thawed at room temperature because any pathogens that may grow during thawing should be killed by the cooking process. Raw meat and poultry can also be thawed under cold running water or in a microwave oven. However, it is still recommended, to follow the 2 hour/4 hour guide in all cases.

Food should be left in its packaging if being thawed under water, to reduce the risk of juices splashing onto other foods, food preparation surfaces, cleaning cloths and staff. The outside of food that is thawing in microwaves generally starts to cook before the food is fully thawed. Care is needed to make sure that food is adequately cooked after microwaving. This is particularly important for poultry, which must be fully cooked throughout. Large portions of meat should also be fully thawed before cooking, otherwise the extended length of time needed to cook from partially thawed may allow pathogens to grow and/or produce toxins to dangerous levels.

It is advisable to thaw raw fish in the fridge or cold room to minimise the chance of production of a chemical toxin called histamine. Histamine, produced by certain spoilage bacteria, is not destroyed by cooking.

One disadvantage of thawing food in a fridge or cold room is the length of time it may take: several days for solid items such as a whole chicken. However, this can be managed effectively if you plan ahead.

Food that is thawing in a fridge or cold room needs to be:

- labelled with the date thawing commenced and identifying information (if removed from original packaging)
- placed in an appropriate container or packaging so any leaking juices cannot contaminate other foods
- if raw, placed below cooked or ready-to-eat foods.

Once you know the thawing times required for different ingredients, prepare written instructions and check they are followed. This removes the risk of undercooking if the food is not fully thawed when it is required.

Food preparation equipment

It is better to use equipment, including utensils and chopping boards, dedicated to specific food types. This is particularly true for raw and ready-to-eat foods, and allergen and non-allergen containing foods. In the case of chopping boards, different coloured boards can be used for the two types. If it is not possible to use separate equipment (e.g. electrical items such as food processors), they should be cleaned, sanitised and thoroughly dried between each new task.

Simply rinsing off any visible food particles is not adequate; see Chapter 3 for recommended equipment cleaning and sanitising practices.

Labelling and keeping track of individual components

As you prepare products, you must keep track of the preparation date and, the batch number of each component of the product (e.g. a filling for pasta and the pasta dough) through each stage. Carefully checking and recording batch codes and storage dates reduces the risk of accidentally using product components that may be beyond their recommended shelf-life, or including a component containing allergens in a product intended to be allergen-free. It also allows you to track where you have used particular ingredients should you ever have to do a recall or if one of your suppliers recalls an ingredient.

Making unplanned changes

You may at times be tempted to alter the way you prepare your products without first using a methodical planning approach. For example, you might wish to experiment and try out new ideas or circumstances may arise that put pressure on you to make changes quickly. These practices are highly inadvisable in a food manufacturing or processing business. Any changes should be planned carefully and their potential impact on food safety fully assessed.

Changing your recipe

If you run out of an ingredient halfway through preparation, it is generally not advisable to leave it out of the recipe or swap it for another ingredient. Remember, you have already taken care to ensure your product meets your specifications, such as the correct pH and/or water activity. You already know which (if any) allergens are present in your product. Changing ingredients or proportions may interfere with your recipe hurdles and also affect the cooking process required to make your product safe.

Changing your normal portion or batch size

During unexpected peak periods, it is not advisable to scale-up your batch sizes without first pausing to check you are not potentially increasing any food safety risks. Increasing batch sizes for products that require specific cooking and cooling procedures, without prior planning, is particularly hazardous and is not recommended (see later in this chapter for more information). For example, you may not have sufficient refrigerated space available to chill large batches, and improvising may lead to temperature abuse and pathogen growth. If you do need to make more product in an 'emergency', it is advisable to stick to the same recipe, using the same ingredients and ingredient quantities, cooking and cooling procedures, but to make multiple batches.

Clearing and cleaning as you go

These practices should become second-nature to you and your staff as you go about your daily work activities:

- Discard outer packaging, such as cardboard boxes, when receiving ingredient deliveries, as it can be a source of contamination in the production area.
- Throw away food waste and inner packaging as soon as possible, taking extra care with drips from raw meat and chicken juices or eggs.
- Clear away small kitchen appliances to the pot wash area as soon as you have finished using them.
- Immediately clean up spills or drips with a disposable cloth, paper towel or clean re-usable cloth.
- Clean down surfaces between each task with a disposable cloth, paper towel or clean re-usable cloth. If handling ready-to-eat food, also sanitise the surface.
- If using one sink to wash raw and ready-to-eat foods, clean and sanitise the sink between uses.

It is important to note that these practices do not replace thorough routine cleaning and sanitation at the end of the day or work period.

Preparing products containing food allergens

If you make multiple products and some contain food allergens while others do not, you must take great care to avoid transferring any of the allergens into the products that are intended to be allergen-free.

Always schedule the order of preparation of products so that allergen-free foods are prepared before allergen-containing foods. All equipment, surfaces and utensils that come into contact with food allergens must be thoroughly cleaned before re-use in products deemed to be free of allergens. Even tiny particles of allergens that are too small to see are dangerous and may cause an allergic reaction in a susceptible individual.

If you prepare a significant amount of allergen-free food and have sufficient room, you should consider setting aside a special production area for these foods. This would also contain a separate set of utensils and equipment.

Cooking – some like it hot, but not pathogens ...

Since man first discovered fire, cooking has been used to improve the texture, flavour, digestibility and storage life of foods. Although foods may need to be 'cooked' (baked, boiled, fried, etc.) so their taste and texture are improved, more importantly they may be 'heat processed' as a way to reduce levels of pathogenic and spoilage microorganisms present on raw foods and ingredients. The conditions for heat processing need to be more strictly controlled and monitored than cooking. This distinction is important for you to understand, because it is an example of one of the key differences between preparing food on a small scale for your family or restaurant compared with commercially manufacturing products on a larger scale.

Sterilising and pasteurising are the two heat processing methods commonly used by food manufacturers. Sterilising usually involves processing at temperatures above 100°C for over 10 minutes and pasteurising usually involves processing at 70–100°C for 10 minutes or less.

Special equipment and technical expertise are required to sterilise foods safely. Sterilisation treatments should not be attempted if you do not have access to these essentials. Pasteurisation, on the other hand, can be done safely by most small businesses.

The heat process that is appropriate for your product depends on the:

- product characteristics, including pH, water activity, and the amount and type of pathogens possibly present in ingredients used
- type of packaging used for the product during the heat process
- intended storage and distribution time of the product (shelf-life)
- intended storage and distribution conditions of the product (chilled, frozen or shelf-stable)
- intended product use; including whether it is to be re-heated before eating or whether it is likely to be eaten by people more vulnerable to foodborne illness. These people include babies and young children, the elderly and people who are ill or have weakened immune systems.

Foods that go through a sterilisation process in appropriate packaging can be safely stored at room temperature (shelf-stable). Foods that are pasteurised generally need to be kept at 5°C or below for a limited time or have some other hurdle(s) applied to allow them to be stored safely at room temperature; for example, reducing their pH, see Box 63 under 'Pasteurisation' for details.

Commercial sterilisation

'Sterilising' in the food industry means heat processing using specific temperature and time combinations sufficient to achieve what is referred to as 'commercial sterility'. Commercial sterility does not necessarily mean that the food is completely free of microorganisms, but that any remaining after processing will not be able to grow under normal storage conditions.

To ensure safety in low-acid (above pH 4.6), high-water-activity (above a_w 0.85) foods, the heat process must be able to kill a very high number of *Clostridium botulinum* spores that may theoretically be present. *C. botulinum* spores are of primary concern due to their very high heat resistance and the deadly toxin this pathogen can produce.

Canning

Canning refers to the process of sealing a food in a package and heating it to achieve commercial sterility. Correctly canned foods can be stored for long periods at room temperature.

If the canning process is not performed properly, there is an increased risk of *C. botulinum* toxin production. It is highly dangerous to attempt to produce canned food without seeking expert advice.

Even though various forms of packaging can be used, including different types of metal cans, glass, plastic or foil pouches, the process is still called 'canning'. For simplicity, all types of packaging containers used for sterilisation are referred to here as 'packs'.

Commercial-scale canning requires use of a retort: an industrial-scale pressure cooker in which water or steam are used to heat the packs. Typical sterilising temperatures range from 110 to 135°C, so the process needs to be performed under pressure. This is because the temperature at which water boils at atmospheric pressure is only 100°C. It would take a very long time to achieve commercial sterility at 100°C, so this is not practical. Additionally, as the air inside the packs will expand rapidly at temperatures above 100°C, counter-balancing external pressure is needed to stop packs from distorting or bursting during processing.

The time–temperature combination needed to achieve commercial sterility is dependent on several factors such as the:

- pH of the product – low-acid foods (above pH 4.6) require a more severe heat process than high-acid foods (below pH 4.6)
- consistency of the product – viscous (thick) foods take longer to heat than very liquid foods
- ratio of solids to liquids in the pack – products with more solids will take longer to heat
- amount of air trapped inside the pack (headspace) – the bigger the headspace the longer it will take to heat the food
- initial temperature of the product in the pack when the heat process starts – the heat process is designed to start at a specified temperature
- type of packaging material used (metal, glass or plastic) – each material will conduct heat at a different rate
- type of heating medium used in the retort – some heat more efficiently than others
- number of packs loaded and their position in the retort – fuller loads may require more time for the retort to reach temperature
- movement of packs during the heat process – for example, packs that are rotated will heat and cool more rapidly
- method used to cool the product post heat processing – some methods allow for faster cooling than others.

You need to be certain that heat penetrates adequately to all parts of all packs and, to do this, heating trials must be run to determine the slowest heating point and if there are any 'cold spots' within the retort (areas that heat slower than others). To achieve this, specialised temperature monitoring devices such as thermocouples are required, which should be able to measure high temperatures with great sensitivity (i.e. plus or minus 0.1°C).

Once you have evidence that the retort heats all packs evenly, the next step is to do several trials so that any variation between runs can be detected and accounted for. The temperature of the slowest heating point of several packs distributed throughout the retort must be measured. If a cold spot in the retort has been detected, there must be a pack at this point.

After the heating trials, mathematical analysis can be used to calculate the time–temperature combination required to safely process your product. This will only work under the conditions of the trials. If you change any of the factors listed above (e.g. the type of packaging material), you may need to repeat the trials to assure safety.

Following heat processing, the food must be protected from recontamination. It is essential that the packaging is robust enough to withstand the temperatures and pressures during retorting. If any seams or seals are faulty and leak, pathogens and spoilage microorganisms can enter the pack. This post-process contamination is most likely to occur when the package is cooling or subject to large temperature fluctuations. Water used to cool the packs must be sanitised so it is virtually free of microorganisms.

It is critically important that while packs are cooling they are not handled in a way that may contaminate the seal or seam areas (e.g. touching with bare hands).

As you can see, determining safe-processing parameters for canned foods and knowing that the process is correctly applied is a critical and specialised procedure. It is very highly undesirable for anybody to attempt to produce canned foods unless they have received advice from an appropriately trained specialist with Approved Persons status. This status is granted by AQIS to someone who has passed a recognised course in thermal processing. Approved Persons have a sound understanding of how to design safe-processing parameters. It is recommended that this professional advice is sought at the early planning stages. See page 278 for information on how to locate an Approved Person.

Ultra High Temperature (UHT)

Ultra High Temperature (UHT) processing coupled with aseptic filling (Chapter 7) can also be used to achieve commercial sterility. It differs from conventional canning in that the product is sterilised outside the final container. Specialised equipment and expertise is required to safely process and pack UHT products.

UHT is primarily used for liquid products such as milk, soup or stock. The process involves heating the product above 135°C for 2–5 seconds then filling into pre-sterilised containers in a sterile environment. Because the product is heated outside the containers, there is more flexibility in the types and size of packaging that may be used.

There are several different types of equipment available for UHT processing. Those that are generically called heat exchangers all operate using the same principle. The product flows over heated surfaces and the heat is transferred indirectly into the product as it passes through the equipment. Direct steam injection into the product is also used for UHT processing. Equipment used for UHT processing must be maintained rigorously because even very minor leaks can cause contamination.

The time–temperature combination required for products varies, with thicker products taking longer to heat. Trials must be performed with expert guidance to ensure the correct processing requirements are chosen.

Measuring the temperature of food

If you intend to pasteurise your products, it is essential to know how to correctly measure the temperature of food. So, before discussing pasteurisation in detail, some facts on temperature measurement will be provided here.

When making a meal at home for your family, you would generally judge how hot or cold it was by touch or sight only; this practice is not adequate for commercial-scale manufacturing of food products.

The Code specifies: A food business must, at food premises where potentially hazardous food is handled, have a temperature measuring device that:

- is readily accessible; and
- can accurately measure the temperature of potentially hazardous food within 1°C (plus or minus).

Source: Standard 3.2.2 Food Safety Practices and General Requirements

Of course, you must make sure that you then use the temperature measuring device (generally a thermometer) when appropriate. The points in a product's preparation that require temperature measurement should be documented on your food preparation flow chart. Additionally, the flow chart should state the time–temperature combination that needs to be achieved. For example, the flow chart showing the chicken salad preparation steps (provided earlier in this chapter) states that the chicken should be cooked until the thickest point reaches 75°C for 30 seconds.

Accurate temperature measurement is essential, because adequate heating and fast cooling are key food safety control measures.

Where temperature should be measured

The slowest heating point must always be used as the spot to take measurements for canned and pasteurised products. The slowest heating point is, as the name suggests, the area that will take the longest to heat up. When you need to time your cooking process to achieve a specific time–temperature combination, the timing should only commence when the slowest heating point reaches the specific temperature desired. For example, start timing when the temperature reaches the 75°C of the '75°C for 30 seconds' combination.

Liquid foods, such as beverages, soups and sauces, are heated by convection. If you watch a pot of water come to the boil, you can actually see the convection currents moving the water as it gets

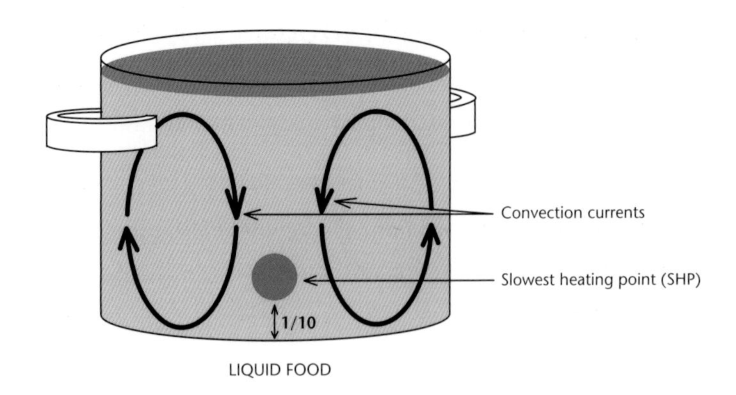

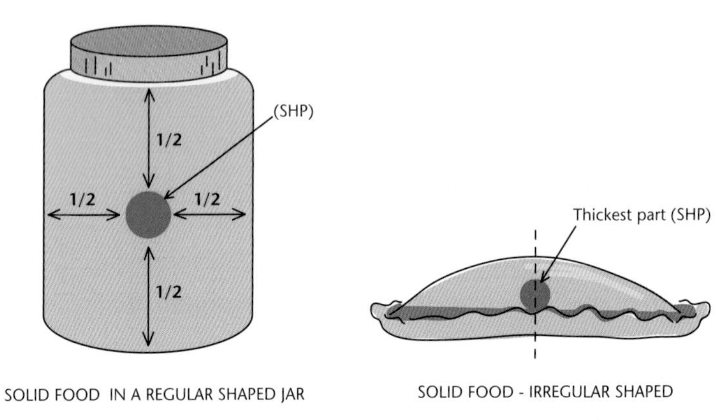

The slowest heating point (SHP) for liquid and solid foods

hotter. These currents move the food around so that hot liquid rises and cooler liquid sinks where it is then heated and rises to the top. The slowest heating point in this situation is about one tenth of the way up the container, in between the convection currents.

In general, the thickest part, or the part in the very centre, is the slowest heating point in solid food such as a piece of meat or a pie. This is because the food particles cannot move around and so heat is transmitted through particles vibrating and warming the particles next to them, which then start to vibrate, and so on. Here, heating starts at the outside of the food and works its way to the centre. This type of heating is called conduction.

The slowest heating point in this situation will be:

- in the centre – for regular shaped foods or those within jars, cans, trays or bottles
- at the thickest part – for irregular shaped foods.

Some foods are heated by a mixture of conduction and convection and the slowest heating point may change as heating progresses; for example, sauces that start off thin but become thick. For these types of foods, measure the temperature at various spots in the product, and at different times throughout the heating process, until you get a feel for where the slowest heating point is for the particular product and heating method.

At other times it may be more important to measure surface temperature, such as when cooking whole cuts of red meat that do not need to be fully cooked through. See Box 60 for information on safely cooking meat. When chilled foods are left at room temperature, it is the surface that warms first so the temperature needs to be measured there, so the food is not left above 5°C for too long.

When temperature should be measured

It is not necessary to measure the temperature of all food items each time they are prepared. This is impractical. The key is to spend some time initially taking many temperature measurements while you are working out your methods. Once you have determined the processing parameters required (e.g. the temperature to set the oven, the pre-warming time and the length of time products should be placed inside) and documented them, you can take readings less frequently. Comprehensive measurements only need to occur again if the method, food or equipment used changes.

Visual checks can also be used; for example, if soup is bubbling assume at least 90°C has been reached. However, it is essential to regularly mix or stir liquid foods during heating.

If you routinely cook multiple items in batches, such as pies in an oven, an acceptable practice would be to determine (by prior testing on multiple occasions) where the slowest heating point in the product is and where the cold spot in the oven is. Then the slowest heating point of the pie located in the oven's cold spot routinely needs testing, because this will take the longest to reach the specified temperature.

Thermometers used during cooking and cooling

There are several different types of thermometers that may be used to monitor the temperature of foods while they are being cooked or cooled.

The essential requirements are that they must be:

- capable of being accurate within 1°C
- suitable to immerse or stick into foods
- able to be effectively cleaned and sanitised (Box 61).

Various types of thermometers are available for measuring the internal temperature of food, such as:

- bimetal or dial probe thermometers
- digital thermometers:
  - thermistor probe
  - thermocouples.

See the photos below of some examples of different thermometer types.

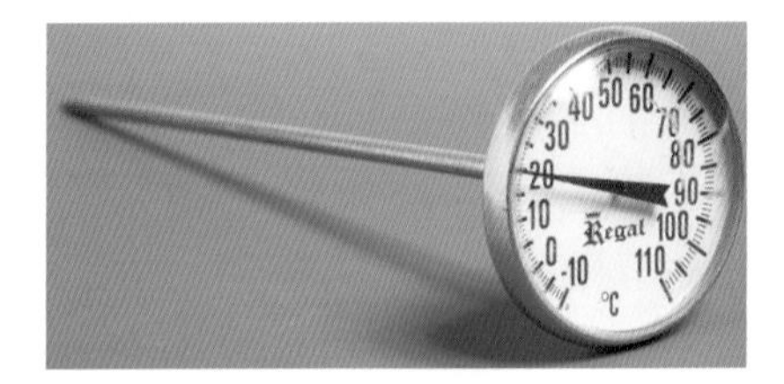

Bimetal probe

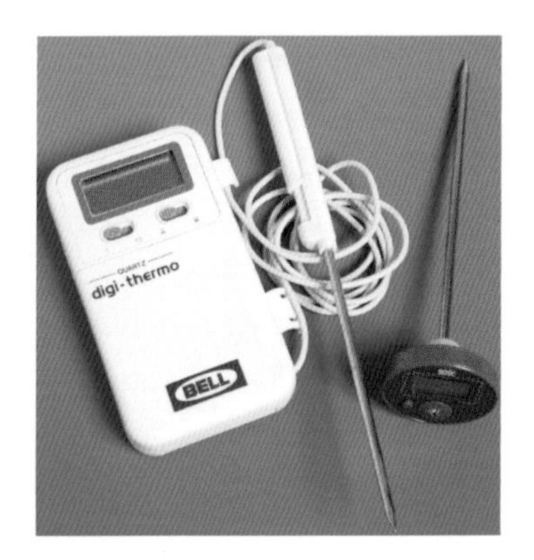

Digital thermistor probes

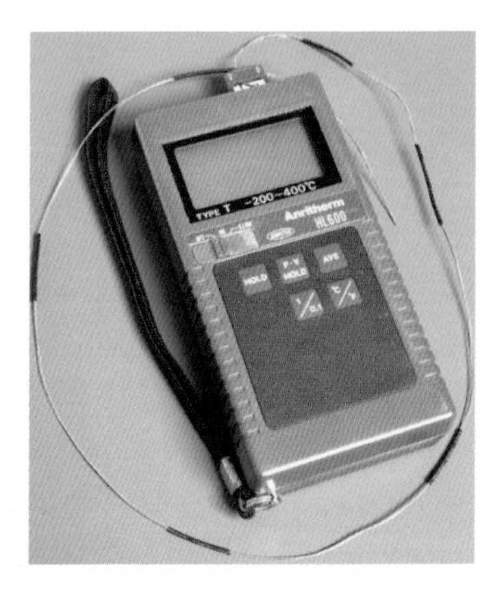

Digital thermocouple

Box 60 – Safely cooking meat

- **Red meat including pork (whole cuts)** – microbial contamination will only occur on the outside of solid pieces of meat such as steak. If you make sure the surface of the meat is well browned, you can cook the centre according to your preference.
- **Fish (whole pieces)** – microbial contamination only occurs on the outer surface, so it can be cooked as per red meat.
- **Whole poultry (e.g. chickens, turkeys and ducks)** – microbial contamination is likely throughout. You must make sure the internal temperature at the slowest heating point reaches at least 75°C. The slowest heating point is usually the thigh or the centre of the cavity if stuffing has been used.

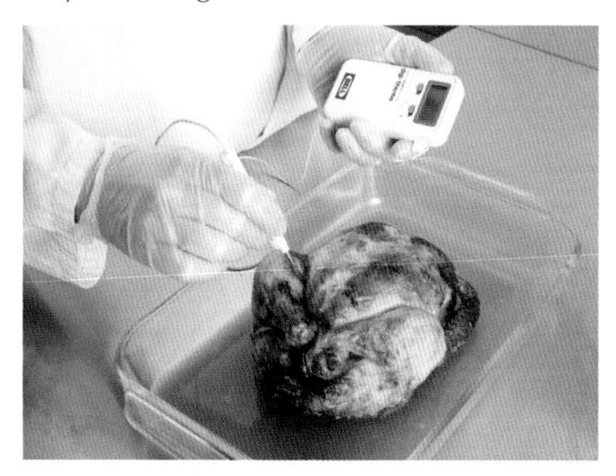

- **Processed meats and other meat products**:
  - minced meat
  - stuffed, rolled or boned meat
  - mechanically tenderised meat, which has had small holes made in its surface that penetrate into the meat
  - corned beef, which has had needles inserted into it to pump in a brine solution
  - sausages

  microbial contamination will occur throughout these items, so you need the internal temperature at the slowest heating point to reach at least 75°C.

The one exception to the above guidance on cooking whole cuts and pieces of meat or fish is if you have made slits or pockets on the surface of meat or fish (e.g. to push herbs into). In this case, you should make sure that the meat or fish is cooked through to the same depth as these slits or pockets.

Digital thermistors give a faster and more accurate reading than bimetal thermometers and so they are more useful where large temperature differences occur. They are also able to measure the temperature of relatively flat foods because only the tip of the probe needs to be immersed.

Bimetal thermometers need to be immersed to the specific depth marked on the probe to get an average temperature reading along the stem (this is usually about 5–6 cm down). Because this is an average temperature and the stem can influence heat transfer, the reading is not as accurate as a digital thermometer and may not be accurate within 1°C. However, some bimetal thermometers have the advantage that they can be left in a food while it is cooking.

Digital thermocouple thermometers are the most accurate and give the fastest reading. For food use, they should be fitted with a type T thermocouple sensor. They can be calibrated, but tend to be more expensive. They also tend to be more bulky than bimetal and thermistor thermometers.

Your EHO and equipment suppliers will be able to advise you on the best type of thermometer to purchase for your situation.

Thermometers need to be regularly maintained and calibrated so they can work accurately (Chapter 3, Box 25). However, if the temperature probe says the food is hot enough but it still does not look cooked or hot then rely on your experience and judgement rather than on the equipment, which may have a fault. For example, if you cannot see steam coming off food it may not have reached the required temperature.

Pasteurisation

The purpose of pasteurisation is to reduce the levels of, or completely kill, pathogens or spoilage microorganisms that are of primary concern in specific products. These are termed target microorganisms; see Box 62.

Because pasteurisation does not kill all pathogens, particularly heat resistant spores, it is essential to use other hurdles to ensure the safety of low-acid foods (above pH 4.6). One of the primary additional hurdles for low-acid foods is to store products at 5°C or below (for a limited time) after pasteurisation. The 'Cook chill' segment below discusses pasteurisation combined with chilled storage.

One of the other additional hurdles is to add acid to low-acid food to change it to a high-acid food, before pasteurising. Once the pH has been changed to below 4.6, the target microorganisms then become spoilage yeasts and moulds because pathogen growth is restricted below pH 4.6. This is the process used during vegetable pickling; see Box 63.

Box 61 – Cleaning and sanitising a thermometer

When using a probe thermometer in foods, cross contamination of pathogens or allergens from one food to another is a potential hazard.

Between uses in the same food type but different batches, a quicker method can be used. You should wipe any food particles off the probe and then wipe the probe with an alcohol wipe to sanitise it between different measurements. Swabs or wipes containing another sanitiser type can also be used.

Between uses in different food types during the day, you should clean and sanitise more thoroughly. This is especially important when measuring the temperature of both raw and ready-to-eat foods, and allergen-containing and allergen-free foods. First immerse the probe in soapy water to wash and then sanitise, using the same procedure for sanitising utensils and equipment.

This thorough procedure should also be used at the end of each day's use, regardless of the food types the probe was used in.

Packaging and pasteurised products

Detailed information on packaging is provided in the next chapter. However, it is important to discuss here how the packaging method you use can affect the safety of pasteurised products.

Food may be filled into its final packaging before pasteurising (termed heating 'in-pack') or it may be packaged after pasteurisation.

Heating in-pack offers the greatest protection against post-process contamination, if used correctly. One of the key requirements is that the packaging is hermetically sealed (i.e. air and watertight) after filling, preventing pathogens from entering the pack and contaminating the product after heating.

Aseptic filling after heat processing also offers high protection against post-process contamination. However, aseptic filling is rarely used for non-liquid, low-acid (above pH 4.6) pasteurised products because it involves filling into pre-sterilised containers in a sterile environment. As this requires specialised equipment and expertise, it is not a common practice to combine with low-acid product pasteurisation, which does not achieve commercial sterility.

Box 62 – What are target microorganisms (for food safety)?

The target microorganism relevant to the safety of a specific product type is the pathogen that is:

- likely to be present in the product
- may cause a food safety issue if not controlled in the product
- generally the most resistant to the hurdles used than other pathogens that are possibly present in the product.

If your heat process and/or other hurdles effectively control the target microorganism, then the other potentially problematic pathogens for your product should also either be killed or prevented from growing to unacceptable levels.

Historically, the target microorganism for pasteurised milk was the pathogenic bacteria *Mycobacterium tuberculosis,* which causes tuberculosis (TB). Nowadays, the pathogenic bacteria *Coxiella burnetii,* which causes Q-fever, is the target microorganism for pasteurised milk.

Other examples of target microorganisms are *Listeria monocytogenes* in cooked cured meats, *Staphylococcus aureus* in low-water-activity foods, and both *Listeria monocytogenes* and *Clostridium botulinum* in low-acid cook chill products.

Hot filling after heat processing is used more often than aseptic filling for pasteurised products. The products are placed into their final packaging while still hot from their pasteurisation process, and then the packs are immediately sealed and inverted (i.e. turned upside down) to allow the heat from the product to pasteurise the top inner packaging surface. If a minimum filling temperature of 85°C is used, the chance of post-process contamination with spoilage microorganisms is greatly reduced.

If the products are high acid (below pH 4.6), then the combination of the low pH, pasteurisation and hot filling is sufficient to control microorganisms if the correct procedures are followed. These products can be stored at room temperature. See Box 63 for more information.

However, if low-acid (above pH 4.6) products are hot filled, there is a chance of post-process contamination with pathogens potentially able to grow in the product. These products should be:

- filled while all parts of the product are above 85°C
- filled under strict hygienic conditions
- stored at 5°C or below for a limited time.

In general, hot filled low-acid products have a maximum shelf-life of 10 days. Rapidly freezing these products would be a preferred option, in which case a long shelf-life can be safely achieved. See 'Freezing post pasteurisation' later in this chapter for details.

Box 63 – Combining pasteurisation and acidification

Low-acid foods (above pH 4.6) such as vegetables can be safely prepared so they can be stored at room temperature, without the need to use a commercial sterility heat process. The method used to achieve this is a form of pickling, and the primary steps are:

- adding acid to change the pH of all components to below 4.6
- pasteurising
- hot filling into jars or heating in-pack.

It is recommended that the pH of foods should be reduced to pH 4 or lower, because this helps reduce the risk of product spoilage. The examples given below are for products at pH 4 or lower.

Hot filling is generally used for thick sauces and purees. The primary steps are:

- heating the product to over 85°C (preferably 90–95°C)
- filling while the product is at least 85°C into pre-warmed glass containers
- immediately securing the lids
- tipping the filled containers upside down and holding for 3 minutes at room temperature.

Heating in-pack is generally used for food pieces in syrup or brine. The primary steps are:

- preheating the product before filling to shorten the time needed for heating in the container (this is not a food safety requirement but may improve quality)
- holding the temperature at the slowest heating point at 85°C (or higher) for at least 2 minutes
- leaving the lids half a turn short of being fully closed during heating, to prevent pressure build-up inside the container distorting the lid

- immediately sealing the lids fully upon removing from the hot water; a special tool can be used to screw down the hot lids, minimising the risk of burning your hands or contaminating the product.

As the product is hot filled into jars, or is heated in jars, there is no need to pre-sterilise the jars: just make sure that they are clean. Never re-use lids as they are not designed to work effectively more than once.

Once you are happy with your product recipe, have decided which methods you will use and have chosen the containers, you must prepare some trial batches. This allows you to work out how to heat your product so that you achieve the time–temperature combinations recommended above. During trials for heated in-pack products, you should measure the temperature in your products, rather than only measuring the temperature of the water they are immersed in. Once you have finalised your method, remember to document it. You do not need to routinely measure your product temperature each time, as long as you don't change any parameters (e.g. recipe, equipment, packaging or heating method).

Cook chill

The term 'cook chill' is used to describe food processing that involves a pasteurisation step, followed by chilled storage before eating. The food should be fully cooked and may only need re-heating by the consumer or may be eaten without any further heating. Examples of cook chill products are pasta sauces, curries, stews, some baby foods, soups and some dips.

Because cook chill products are generally low acid (above pH 4.6) and high water activity (above a_w 0.85), the product recipes can't be relied upon to prevent the growth of pathogens to dangerously high levels. Therefore most cook chill products are potentially hazardous foods. For some examples of foods that are potentially hazardous see Chapter 3, Box 22.

If you are producing potentially hazardous cook chill products, you must closely follow the pasteurisation, packaging, handling and shelf-life guidelines recommended here to reduce the chance of causing foodborne illness.

Pasteurisation, packaging and storage guidelines for low-acid cook chill products

Pasteurisation processes	Packaging method	Shelf-life at 5°C or below	Important requirements
Note: – these are minimum requirements – temperature must be measured at the slowest heating point			
70°C for 2 minutes or equivalent time–temperature combination (see table on page 185)	Heated in-pack or aseptically filled	10 days maximum	Store at 5°C or below at all times
70°C for 2 minutes or equivalent time–temperature combination (see table on page 185)	Hot filled	10 days maximum	Minimum fill temperature of 85°C. Store at 5°C or below at all times
90°C for 10 minutes or equivalent time–temperature combination (see table on page 185)	Heated in-pack or aseptically filled	Over 10 days – to upper limit advised by technical expert	Technical expert guidance strongly recommended. Store at 5°C or below at all times
90°C for 10 minutes or equivalent time–temperature combination (see table on page 185)	Hot filled	10 days maximum – unless a technical expert advises otherwise	Minimum fill temperature of 85°C. Store at 5°C or below at all times

If you use the 70°C for 2 minutes (or equivalent) pasteurisation process, the maximum shelf-life you can safely achieve for low-acid products is 10 days. This is because this process will only control bacterial cells and not pathogenic spores. If you choose to pasteurise your products using the 90°C for 10 minutes (or equivalent) option, it is possible to safely achieve a shelf-life of greater than 10 days. However, you must be certain you are heat processing correctly; see under 'Time–temperature combinations: do you need expert guidance?' (page 184) for further information.

You may wish to further enhance the safety of products heated at 90°C for 10 minutes, if, for example, you lack confidence that a storage temperature of 5°C or below can be maintained.

To do this, additional hurdles can be used:

- pH – adjusting to pH 5 or below
- Water activity – adjusting to 0.97 or below.

Alternatively, you may choose to freeze your products so they are 'cook freeze' instead of cook chill. See under 'Freezing post pasteurisation' (page 192) for more information.

If the overall pH of your product is well below pH 4.6 (high acid), your product is not able to support the growth of pathogenic microorganisms and can generally be cooked to your liking. However, if your product contains meat, you will probably not be able to achieve a pH below 4.6 throughout.

Detailed guidance on the safe production of cook chill foods is provided in Cook chill for foodservice and manufacturing: Guidelines for safe production, storage and distribution (see page 271 for details).

Time–temperature combinations: do you need expert guidance?

If you are producing cook chill potentially hazardous products, it is essential that you achieve the recommended time–temperature combination relevant to your desired shelf-life. This must be achieved at the slowest heating point of the product. For example, if using the 90°C for 10 minutes pasteurisation process, the total time required will be more than 10 minutes because you need to wait until the slowest heating point reaches 90°C before starting to time the 10 minutes.

Like canning, many variables affect the time–temperature combinations required to pasteurise foods adequately:

- **Batch size** – if you change batch sizes of your product you must re-test and adjust your heating time.
- **Solid-to-liquid ratio of your product** – liquids heat faster than solids so if this proportion changes you must re-test and adjust your heating time.
- **Temperature of the food at the start of processing** – if the food is chilled at the start of the pasteurisation, it will need a longer time to reach the desired temperature, compared with warmer food. You must use the same starting temperature that was used when your heat processing method was first designed, otherwise you will need to redesign and test your process.

Equivalent time–temperature combinations: 70°C for 2 minutes

Temperature	Time
Note: – these are minimum requirements – temperature must be measured at the slowest heating point	
60°C	45 minutes
65°C	10 minutes
67°C	5 minutes
72°C	1 minute
75°C	30 seconds
78°C	10 seconds
85°C	1 second

Source: adapted from Cook chill for foodservice and manufacturing: Guidelines for safe production, storage and distribution (see page 271 for details).

Equivalent time–temperature combinations: 90°C for 10 minutes

Temperature	Time
Note: – these are minimum requirements – temperature must be measured at the slowest heating point	
78°C	3 hours and 40 minutes
80°C	2 hours and 10 minutes
82°C	1 hour and 20 minutes
84°C	47 minutes
86°C	28 minutes
88°C	17 minutes

Source: adapted from Cook chill for foodservice and manufacturing: Guidelines for safe production, storage and distribution (see page 271 for details).

- **Heating method used** – whether it is immersed in water, steamed or heated directly will affect how quickly heat is transferred through the product, and the time–temperature process required.

Take note that that the entire product (i.e. every single ingredient) must receive the minimum required time–temperature combination. If you add ingredients at different times of the cooking process then thorough mixing is required, followed by re-heating to the desired temperature (at the slowest heating point) and starting the timing process again. See Box 64 for an example of this scenario.

If heating in-pack, the additional variables that will affect the time–temperature combination your product requires are:

- the weight or volume of the package
- the headspace or airspace in the package
- the type of packaging material used; for example, how thick it is and what is it made of
- the shape of the filled pack; for example, thin packs heat faster than thick or cylindrical packs.

If any of these factors change, you must re-test and adjust your temperature and/or heating time accordingly.

Many standard cooking processes would generally achieve the 70°C for 2 minutes or equivalent time–temperature combination. However, it is still important that you check this by measuring the temperature at the slowest heating point for several batches. This should preferably be done on several different days so any variations can be detected. Once you are satisfied that the method you are using is achieving the desired combination, you can then decrease the frequency of these temperature checks.

Multiple batches should be checked again if variables are changed, including:

- the heating method
- the recipe
- the size of ingredient pieces
- the amount of sauce or liquid
- the volume/weight of batches or individual packs
- the equipment
- the packaging
- the starting temperature of the product.

Box 64 – Multi-stage pasteurisation

A soup is cooked by bringing to the boil and then simmering for 15 minutes, exceeding the requirement for the 90°C for 10 minute time–temperature combination. However, the temperature of the soup is then allowed to cool down to 60°C before adding in cream and parsley. After these additions are made, the soup is re-heated until it reaches 70°C (at the slowest heating point) and then it is held for 2 minutes. This now means that the soup should be stored for up to 10 days only (unless other hurdles are also used).

If you want a product to have a shelf-life of greater than 10 days, it is extremely important that you correctly and consistently achieve the 90°C for 10 minutes (or equivalent) time–temperature combination for the slowest heating point within each pack. It is strongly recommended that you have a technical expert, such as an AQIS Approved Person, validate your heat process.

Validation of a heat process involves obtaining numerous time and temperature measurements, usually using thermocouples with a high degree of accuracy. Results from these trials are used to determine the total heating time required to achieve a heat process equivalent to 90°C for 10 minutes. You should also arrange for the technical expert to:

- check for any hot or cold spots in your heat processing equipment that may lead to uneven heating
- advise you on what to do if a batch is under processed
- determine the heat processes required if the product is at a different starting temperature than normal
- advise you on how to perform your in-house routine temperature checks including:
  - where the products slowest heating point is
  - what type of temperature measuring device should be used
  - the location of any cold spots in your equipment
  - how frequently batches should be checked
  - how many individual packs should be checked each time
  - how much variation is acceptable.

Cooling – now your product is hot you must cool it down ... carefully

In this segment, cooling requirements for pasteurised products only will be discussed. The method for safely cooling products processed to achieve commercial sterility will be determined by the technical expert you engage to develop your canning or UHT processing parameters.

Some pathogens can survive the pasteurisation heat processes. Spore-forming bacteria, such as *Clostridium perfringens*, not only survive but may be activated from their dormant state so they are capable of growth. It is for this reason that is essential to cool food down quickly after cooking.

Timeframes for cooling food after cooking are specified in the Code (Standard 3.2.2 Food Safety Practices and General Requirements).

Food must be:

- **cooled to 21°C within 2 hours, and then**
- **cooled to 5°C (from 21°C) within 4 hours.**

These timeframes start from when the food cools down to 60°C, because pathogens cannot grow above this temperature. The total time for cooling from 60°C to 5°C (or below) must be no more than 6 hours.

Initially, you must determine whether your cooling method is meeting these requirements. If not, you must adjust your method until you can consistently achieve them, unless you can show that your method will not allow pathogen growth or toxin production to unacceptable levels. For example, Safe Food Australia provides alternative safe methods for cooling uncured and cured meat products that may be difficult to cool quickly, depending on their size and shape.

You must go through this process for all the different pasteurised product types you make because, as with heating, many variables will affect the way a food cools down. Once you have determined the correct method for cooling products, you should document the details in your product-preparation flow charts. If you make any changes to your products, or their heating and cooling methods, you must re-test to see if you are still achieving the requirements of the Code.

It is essential that the speed at which you prepare batches of pasteurised products does not 'out run' your chilling capacity. If you have products sitting at room temperature for longer than about 30 minutes after heat processing, it is advisable to slow down your production rate or purchase more chilling equipment.

Reducing portion sizes before chilling

Splitting up hot food into smaller portion sizes will reduce the time required to chill the food. Ideally, this should occur when the food is still steaming hot, because this will reduce the risk of contaminating the food while handling. Also, to reduce the risk of contamination, this process should occur in an area that has been thoroughly cleaned and sanitised and is not near to any other food preparation activities (particularly handling of raw foods) or cleaning operations.

Box 65 shows a real-life example of how slowly a large pot of soup takes to cool down at room temperature if not split into smaller portions. Clearly, this does not meet the cooling requirements of the Code.

Chilling equipment

In addition to reducing portion sizes, cooked products will need to be transferred into chilling equipment for cooling. Make sure you don't exceed the maximum recommended capacity for the equipment, and follow the manufacturer's instructions for procedures for loading with food and equipment maintenance. Cleaning and sanitation procedures for chilling equipment must be strictly followed so *Listeria monocytogenes*, which likes damp and cool conditions, does not 'take up residence' within the equipment.

Blast chillers

Blast chillers use powerful fans to circulate cold air around food, speeding up chilling considerably (compared with standard refrigeration). Using a blast chiller allows you to cool larger amounts of foods in larger portion sizes, while still meeting the requirements of the Code.

Foods may be left uncovered while in thc blast chiller, allowing the process to speed up further. Care is required to clean and sanitise the equipment in between uses to reduce the risk of contamination. Because the food is placed on racks or shelves, blast chillers are ideal to use if your products are fragile and prone to damage, such as pies or cooked vegetables.

Cold store rooms

If portion sizes are kept small, a normal cold store room – fitted with a fan for air circulation – may be used as long as the cooling requirements of the Code are met. Packs of food should be spread out to allow the cold air to flow around them effectively.

Box 65 – Soup chilling

A trial was conducted to see how long it would take a large pot of thick soup to cool down to 21°C (from 60°C) if it was left to chill at room temperature in the pot in which it was cooked.

Details of trial:

- Product: 6 L of thick lentil soup
- Pot: 7 L stainless steel
- Room temperature: 12–14°C
- First stage cooling: 100°C down to 60°C = 3 hours (there will be no growth above 60°C so this time is not counted)

Cooling time: 60°C to 21°C

Time (hours)	Temperature of soup
0	60°C
1	50°C
2	43°C
3	38°C
4	32°C
5	31°C
6	29°C
7	27°C
8	24°C
9	22°C
10	21°C

Should be 21°C by now

Should be 5°C or below by now

At the time the trial was stopped, the soup had already been in the temperature danger zone for 10 hours. If the trial was left to continue, it may have still taken another 10 hours or more for it to reach room temperature.

Water baths

Water baths allow for effective chilling of packaged foods. Packages can be immersed in chilled water or brine solutions, or iced water. It is important that the water or ice used is 'food grade' and preferably contains an appropriate sanitiser to prevent contamination in case packs are not completely sealed. You may set up your own manual system using a large vessel containing iced water; this will require that the packs are regularly turned and additional ice is added as required. Mechanical or semi-automated systems, although more expensive to purchase, allow you to have better control over the cooling process and are less labour intensive.

Chilling in appropriate containers

Chilling of bulk foods such as casseroles, rice, soups and sauces should occur in large, flat trays. This increases the surface area of the food exposed to the chilled conditions, speeding up the process. Ideally, the depth of the foods in these containers should be no more than 2.5 cm (25 mm) for a standard fridge or cold room, or up to 5 cm (50 mm) for a blast chiller.

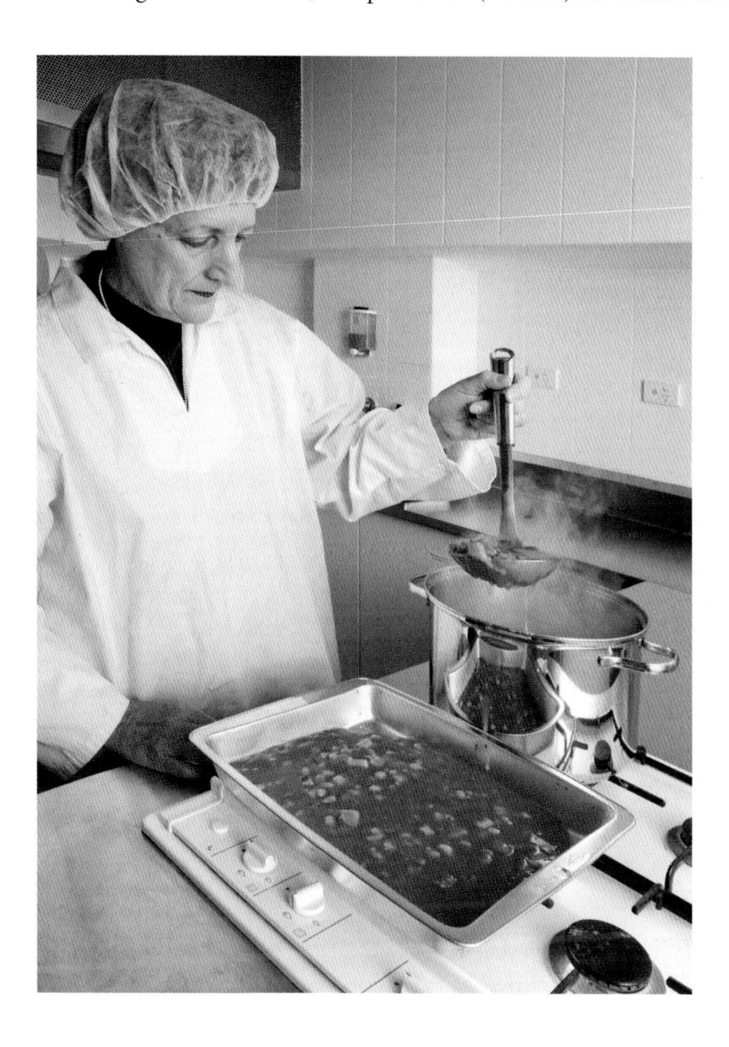

Unless a blast chiller is used, food should always be covered during chilling to reduce the risk of cross contamination.

Metal trays and containers are best used for cooling foods rather than containers made from thick plastic because the plastic acts as an insulator. In addition, it is generally easier to clean and sanitise metal containers compared with plastic.

Freezing post pasteurisation

Freezing low-acid pasteurised foods offers a significant food safety advantage over chilling because pathogenic bacteria are unable to grow in food stored below minus 5°C. If your heat process and packaging combination only allow you a 10 day maximum shelf-life (and you want longer), or you can't be certain you can maintain chilled storage at 5°C or below, freezing your product is a safe solution. For the optimum quality and extended shelf-life, minus 18°C is the recommended temperature.

It is important to note that the freezing process does not kill all pathogenic microorganisms. Although some cells may die, pathogens present in your product before freezing will remain throughout storage. Therefore, you should apply the same food safety controls to frozen foods as you would to chilled foods. It is also essential that you label your products with instructions for cooking or re-heating from frozen. Guidance for this will be provided in Chapter 7.

It is important to make sure you use the appropriate type of packaging for frozen foods to prevent moisture loss (dehydration). Packaging suppliers should be able to tell you which packaging materials provide suitable barrier properties for frozen foods.

Blast freezers are a good equipment choice for freezing on a scale suitable for small businesses. Domestic freezers are not designed to freeze large quantities of food. If you do use a domestic freezer, products should be pre-chilled to below 5°C before freezing to speed up the process and minimise the chance of growth of microorganisms. Depending on the specific equipment capabilities and the volume of product made, you may also need to pre-chill your products before loading into a blast freezer.

Remember, if you are cooking before freezing (i.e. making cook freeze products), you have to comply with the cooling requirements specified in the Code.

Chilling without a prior cooking step

Low-acid (above pH 4.6) products that are not cooked at all before being eaten are among the highest risk products regarding pathogenic microorganisms. Included in this category are salads, including fruit salad, and some uncooked fresh dips.

It is essential that strict hygiene is used when preparing and packaging these products because there is no kill-step to eliminate introduced pathogens. Even sanitising fresh produce using the method recommended earlier in this chapter may only reduce pathogen numbers very slightly.

You also need to be even more rigorous with refrigeration: keeping ingredients and final products at 5°C or below as much as practicable. Raw foods (such as meat) must be kept well separated from these foods to reduce the risk from cross contamination. Use-by dates on ingredients must be followed and appropriate use-by dates for finished products must be determined. See under 'Product shelf-life and food safety' in Chapter 7 for further information.

Producers of vegetable-based dips or mayonnaise-style products (including aioli) are advised to reduce the pH of their products to 4.6 or below (Chapter 4, Boxes 40 and 42).

Microbiological testing provides important evidence

Setting up a microbiological testing program is strongly recommended if you produce potentially hazardous cook chill products or low-acid products that are not cooked (i.e. have not been heat processed).

FSANZ has developed guidelines for the acceptable microbiological quality of ready-to-eat foods, Guidelines for the microbiological examination of ready-to-eat foods (see page 272). Although following these is not mandatory – unlike the microbiological criteria specified for certain product types in the Code (Standard 1.6.1) – it is advisable.

For example, you may choose to do initial testing on several batches of a potentially hazardous raw ingredient purchased from a new supplier (or preferably ask the supplier to show you test results they have). Then, once you are satisfied that the results show no food safety issues, you could reduce the number of tests or results reviewed. You may also choose to do testing for specific target pathogens in your final products, at the end of their shelf-life. Over time, the frequency and number of samples tested can be scaled down, unless you make changes to variables such as raw ingredient suppliers or your heat processing method.

If you produce ready-to-eat products, it is also a good idea to do regular microbiological testing to check for any hygiene issues. In this case, both your final products and food contact surfaces (using swabs) can be tested for the total number of bacteria present. A rise in levels over time indicates something is amiss (e.g. your cleaning and sanitising program may not be effective).

Because specific expertise is required to perform these tests and interpret the results, a commercial testing laboratory should be used. Unfortunately, this is not always a cheap exercise, but if potential problems are detected early, it may save you a great deal of money later on.

KEY MESSAGES FROM CHAPTER 6

- Preparing written plans outlining product preparation steps helps to identify where specific food safety controls are needed.
- Making unplanned changes to recipes, preparation or packaging methods can be hazardous.
- Pasteurisation treatments allowing a maximum shelf-life of 10 days (e.g. 70°C for 2 minutes), must be used with other food safety controls (e.g. storage at 5°C or below).
- Pasteurisation treatments allowing a shelf-life greater than 10 days (e.g. 90°C for 10 minutes) generally need to be set up by a technical expert, and must be used with other food safety controls (e.g. storage at 5°C or below).
- Commercial sterility treatments (heating over 100°C) must not be used without expert guidance, and can allow foods to be stored at room temperature for long periods (if in appropriate packaging).
- The temperature of food should be measured at the slowest heating point, as this will take the longest to heat to the required temperature.
- Businesses handling potentially hazardous foods must have a thermometer that is accurate within 1°C (plus or minus).
- Cook chill products can be hazardous if not prepared and stored using effective and strictly monitored food safety controls (e.g. storage at 5°C or below for a limited time). If you can't guarantee these controls, then cook freeze may be a better alternative.
- Cooked foods must be cooled, from 60°C (or above):
  - to 21°C within 2 hours, and then
  - to 5°C or below within 4 hours

  The total time for cooling to 5°C (or below) must be 6 hours or less. These are requirements of the Code.
- Strict hygiene must be used in the preparation and packaging of low-acid foods (above pH 4.6) that will receive no kill-step such as pasteurisation.
- Setting up a microbiological testing program is recommended for high-risk products.

October 2009
Tuesday
Wednesday
Thursday
Friday
Saturday
April
August
SUNDAY
MONDAY
TUESDAY
WEDNESDAY
THURSDAY

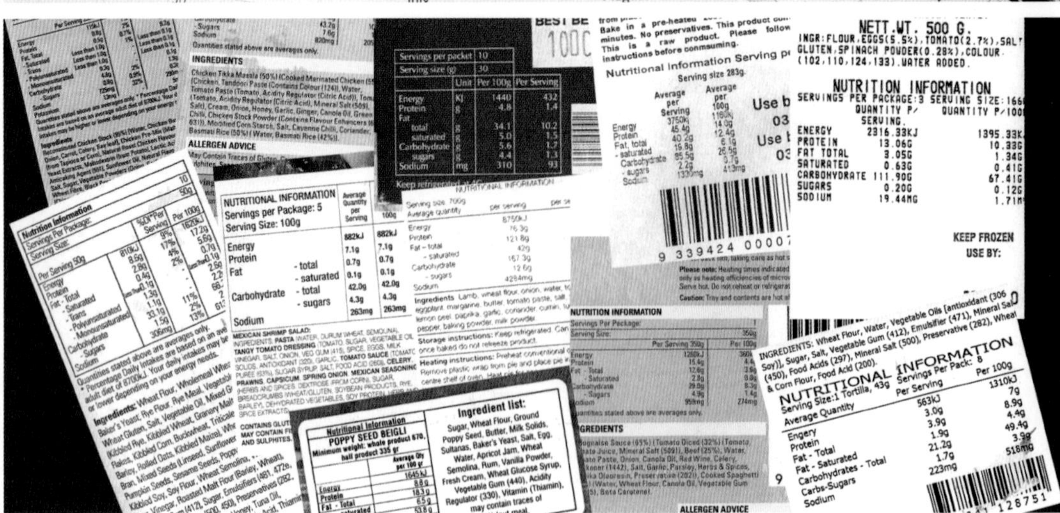

INGREDIENTS
ALLERGEN ADVICE
NUTRITIONAL INFORMATION
Servings per Package: 5
Serving Size: 100g
Nutritional information Serving per
NETT.WT. 500 G.
NUTRITION INFORMATION
KEEP FROZEN
USE BY:
Ingredient list:
POPPY SEED BEIGLI
NUTRITIONAL INFORMATION

Chapter 7

Controlling food safety hazards – packaging, shelf-life and labelling

No matter how well you have applied the food safety guidelines provided here so far, you still risk producing unsafe food unless you make the correct choices for your product packaging, labelling and shelf-life (for products requiring a use-by date). Additionally, if you are producing potentially hazardous foods, you must have adequate temperature control throughout the packaging, storage and distribution stages. The Code (Standard 3.2.2) requires that these foods be maintained at 5°C or below, or another temperature for a specified time if it can be shown this will not allow pathogens to grow or produce toxins to dangerous levels. Specific information on storage and transport of potentially hazardous food has previously been provided in Chapters 3 and 5.

Packaging your products

A major hazard caused by poor packaging choices is the failure to properly prevent microbial contamination, usually through faulty seals. In low-acid (above pH 4.6), high-water-activity (above a_w 0.85) foods, this failure has resulted in deaths and special care is required to ensure packaging remains effectively sealed through the food chain to the consumer.

Packaging is designed to protect, preserve and promote the foods you make. Packaging must protect your food from:

- contamination by pathogenic microorganisms
- attack by vermin such as insects and rodents
- tampering
- undesired chemicals, oxygen and water vapour
- transport damage such as vibration, shock or compression.

Packaging also plays an important role in communicating essential nutritional and dietary information to the consumer, and to 'encourage' consumers to buy specific products.

The information provided here will focus on the factors required to ensure food remains safe and wholesome through the supply chain.

Packaging materials

The range of packaging materials available for food contact use has evolved considerably over time. Where previously rigid glass or steel containers, and paper or cardboard were commonly used, the introduction of plastic as a food packaging material greatly expanded the available choices and allowed for flexible packaging to be developed.

The following table provides you with some information on the different properties and uses of packaging materials. You can use this as background information when discussing your needs with a packaging supplier.

Properties of packaging materials (relevant to food safety)

Material	Packaging forms	Properties
Glass	Bottles, jars	Excellent air and moisture barrier (if sealed correctly) Minimal interaction with food Brittle – hazard if broken (Box 66) Not suitable for frozen foods or rough handling
Metal	Cans, lids	Excellent air and moisture barrier (must be sealed correctly, technical expert required to test can seals) Resistant to high temperatures and breakage Reacts with foods (especially high acid), internal coatings often required
Plastic (Box 67)	Pouches, films, trays, bottles, jars, tubs, sachets	Can be an excellent air and moisture barrier Can be selected for use over a wide range of temperatures (e.g. heat processing for 'commercial sterility' through to freezing)
Paperboard	Cartons, tubs, sachets, pouches	Poor barriers to air, moisture and physical damage unless laminated with plastic and/or foil Can serve as extra physical protection (as secondary packaging)

Box 66 – Glass jar and bottle safety

Pieces of glass are very hard to detect in food, and can cause serious injury if eaten. Specific food safety controls are required when glass is used for packaging.

The risk of glass breakage due to thermal shock should be reduced. Thermal shock occurs when hot food is filled into cold glass containers or when glass is heated or cooled quickly. Containers that will be filled with hot product need to be pre-warmed. Ask your packaging supplier for specific advice for your products and processing methods.

Glass can also shatter if containers are overfilled. As a general rule, you should have a headspace – the empty space between the top of the food and the lid – of at least 5%. However, too much headspace can cause containers to explode during heat processing. It is essential that you get specific instructions from your packaging supplier.

The following is an example of the procedure required if a glass container breaks during automatic filling on a processing line. Breakages can occur because of misaligned filling heads or because containers come into contact with each other with too much force.

The basic steps required are:

1. Stop the processing line and identify areas where pieces of glass may be.
2. Remove and discard all obvious pieces of broken glass.
3. Discard any containers around the breakage point (see note below).
4. Remove any other containers that may hinder the view of pieces of broken glass.
5. Collect all smaller pieces and fragments of broken glass – look on areas such as ledges, belts and the underside of filling valves.
6. If necessary, wash down the area with a low-pressure hose.
7. Dispense product from the filling head to flush out any glass.
8. Do not re-start the line until the cause has been investigated and, if necessary, fixed.
9. Record the details of the affected batch in a glass breakage log.

Note – the number of containers around the breakage point that need discarding depends on the severity of the breakage. For example, a small chip on a container may only require the containers immediately following the damaged container to be discarded. The complete shattering of a container may require any containers that were open and within a 3 metre radius of the accident to be discarded.

Box 67 – Plastic packaging materials

There is a large variety of plastic packaging materials available; those most commonly used for food and their typical uses are:

- high density polyethylene (HDPE), which is used for bottles for large volume non-carbonated products such as milk, and tubs for salads
- low density polyethylene (LDPE), which is used for films and bags
- linear low-density polyethylene (LLDPE), which is used for stretch or cling wraps
- polypropylene (PP), which is used for tubs (e.g. yoghurt), microwaveable packaging and bottles for hot fill products
- polyethylene terephthalate (PET), which is used for carbonated soft drink bottles and conventional or microwave oven heated meal trays
- nylon, which is used for vacuum packs, processed meats, boil-in-bag and cook-in-bag packs (has similar properties to PET)
- polystyrene (PS), which is used for foam boxes, preformed trays, insulated cups and bowls (e.g. instant noodles); PS should not be heated above 95°C
- polyvinyl chloride (PVC), which is used for trays; PVC should not be heated above 80°C.

Purchasing packaging

Now you should have some idea of what type of packaging materials may be suited to your products and you may have a particular preference for one. However, you might not be able to use your first packaging choice because it is difficult to purchase on a small-scale or expensive specialised equipment is required. You may need to wait until you are in a position to scale-up your operation before you can use your original packaging concept. Additionally, because product packaging, product type and a product's shelf-life are 'interactive', if you alter a product recipe, you may need to change your packaging.

'Fit-for-purpose' packaging

The Code (Standard 3.2.2) specifies that '… a food business must, when packaging food, only use packaging material that is fit for its intended use'. One of the reasons for this requirement is to reduce the risk of non food contact materials being used, which puts food at risk of chemical contamination. If you use packaging material suitable for its intended use and follow all instructions provided by the supplier, this should not be an issue. For example, if your product requires heating in a conventional or microwave oven, the packaging must be able to be safely heated under these conditions. To do this, you would inform your packaging supplier of this requirement and request them to provide you with documentation proving that the packaging material they intend to supply is suitable.

You should also set up a series of trials with your intended product–packaging combination to make sure it is appropriate. The adequacy of the seals and the ability to achieve your desired shelf-life are two things that would need to be tested; more information is given on these later in this chapter.

Packaging suppliers

A packaging supplier should be able to tell you which packaging is suitable for your product type, processing method and intended consumer use. They should also be able to provide you with samples and information about sealing methods appropriate to the packaging. You may also wish to seek the opinion of a packaging equipment supplier or another technical expert.

Some of the questions suppliers may ask you about your food products:

- What is the composition of the product (e.g. is it high fat or acidic)?
- Is the product to be heated in-pack, aseptically packaged or hot filled?
- What is the heat process you intend to use (e.g. pasteurisation or commercial sterilisation)?
- What is the storage temperature of the product (e.g. room temperature, chilled or frozen)?
- Will the product be cooked or re-heated by consumers? Will this be in a conventional oven or microwave oven? What is the maximum temperature that will be reached?
- Do you intend for the consumer to reseal the container after opening?
- Do you require the packaging to be transparent (i.e. so the contents are visible)?
- Is the product soft or does it have hard edges that may puncture plastic packaging?

Details of the type and quality of packaging you receive from your supplier should form part of your approved supplier program (see Chapter 5), which should include product specifications for packaging materials. Specifications will include details of any special properties the packaging material needs to have (such as heat resistance or a barrier to oxygen). You may ask for laboratory test results demonstrating the effectiveness of the material to achieve this.

Receipt of packaging deliveries

Each batch of packaging should be examined to check it meets the requirements of your product specifications. Additionally, you must be able to trace every batch of packaging material you use.

It is strongly recommended that you keep samples of batches of packaging for at least the shelf-life of the products the batches were used for. If a potential problem with the packaging is later discovered, this allows you to determine which product batch needs to be recalled.

Using a form like the one shown in Box 68 will also assist you to have a first-in-first-out system for your packaging materials. It can also be used to record details of packaging that you reject because of issues detected after you commenced use.

Caring for packaging materials

All packaging materials should be handled with care to reduce the risk of damage. Open cartons of plastic packaging material with knives or box cutters carefully to avoid cutting too deeply. Do not stack plastic pouches in a way that may cause creases or interfere with the processing of the food. Large plastic rolls, for example, should not be stored on the ground where the pressure may cause the film to stick to itself. Empty metal cans should be protected from physical damage because this may result in ineffective sealing. Glass containers should be moved gently to reduce the risk of breaking or cracking. Items such as plastic closures and tabs from ring-pull cans may be hazardous if they come loose and contaminate food, so packaging containing these must be protected from accidental damage.

Packaging has to be protected from microbial, chemical or physical contamination before use. It should be kept covered and off the floor to reduce the chance of contamination by vermin, insects or other pests. It should be stored well away from any sources of moisture, including spray from hoses during cleaning and sanitising, to reduce the risk of microbial contamination. Open jars, bottles and other containers should be left within their outer packaging until just before use. Once unpacked, ideally, they should be stored upside down to reduce the risk of

Box 68 – Example packaging material receipt and stock rotation form

SMITH & SONS

PACKAGING MATERIAL RECEIPT FORM AND STOCK ROTATION RECORD

Packaging materials received:

Polypropylene trays (500 mL), polypropylene film roll (500 mm × 100 μm), retort laminate pouch (500 mL) – supplied by Shuys Packaging Pty Ltd

Criteria for acceptance of delivery:

Outer packaging is intact with no signs of water damage or physical damage – if damage is observed, open outer packaging and check contents; reject delivery if inner contents are damaged.

Film					
Date received	**Received by**	**Batch number**	**Inspection and rejection notes:**	**Date started use**	**Date completed use**
2/9/08	*Brian*	*134765-7*	*Delivered after business hours and left out in rain, box soaked*	*NA (not used)*	*NA*
3/9/08	*Brian*	*134765-9*	*OK*	*7/9/08*	

Trays					
Date received	**Received by**	**Batch number**	**Inspection and rejection notes:**	**Date started use**	**Date completed use**
21/9/08	*Brian*	*128546-9*	*OK*	*30/9/08*	*30/10/08*
23/10/08	*Mary*	*128789 4*	*Caused blockage on the line, did not meet size specification, send back to supplier*	*31/10/08*	*NA*

physical and microbial contamination. Plastic packaging should not be exposed to high temperatures, which can cause it to 'breakdown' chemically.

Packaging left over from any production run should be returned to its storage area to reduce the chance that it will become contaminated (e.g. during cleaning of equipment).

If you use multiple types of packaging materials that look similar to each other, you must clearly label each type to reduce the risk of accidentally using the wrong type for a particular product; for example, avoiding the use of a pouch for a microwaveable product that is not made of heat resistant plastic.

Food packaging materials or containers should never be used for any other purpose, such as carrying or storing equipment parts.

Temperature control during packaging

It is essential that potentially hazardous food is not left too long in the temperature danger zone (i.e. between 5 and 60°C) during packaging or filling. This applies to all potentially hazardous foods, regardless of whether a heat process will be used.

Potentially hazardous foods that are heat processed must be cooled down to 5°C within 6 hours (Chapter 6). This is a requirement of the Code (Standard 3.2.2). Potentially hazardous food that is not heat processed should not be left in the temperature danger zone for more than 4 hours.

These timeframe restrictions have to be factored in when determining what your packaging method will be. If you cannot comply, you must alter your practices, such as:

- reducing your batch size
- simplifying your packaging or
- hiring extra staff

unless you can prove that the safety of the food is not put at risk.

Controlling the atmosphere around food

The atmosphere inside product packaging can be altered to improve the shelf-life of food or to enhance its quality. Two methods that are used are vacuum packaging and modified-atmosphere packaging.

Vacuum packaging

As the name suggests, vacuum packaging involves extracting the majority of air from around food. The main purpose of vacuum packaging is to exclude oxygen, making it more difficult for some spoilage microorganisms to grow and/or to reduce chemical changes (e.g. oxidation) that may limit shelf-life. The types of products that are vacuum packaged include dry products such as tortillas, cured and processed meats, cheese, marinated olives and cook chill foods. Food can be vacuum packed either before or after heat processing.

Flexible packaging material is used, which can cling tightly to the surface of the food when the air is removed. The material has to be an excellent barrier to air, both across its surface and at the seals to prevent air from entering. Specialised equipment that first removes the air and then heat seals the pack is required. Vacuum packaging equipment suppliers can provide advice on the type of packaging material suitable for your products.

Rather than enhancing the safety of products such as cook chill foods, using vacuum packaging with these foods without the proper controls in place may create a food safety issue. This is because pathogens such as *Clostridium botulinum* prefer to grow in the absence of oxygen. See Box 69 for further information.

Modified-atmosphere packaging

Modified-atmosphere packaging (MAP) involves replacing the air surrounding a product with a single gas (e.g. carbon dioxide) or mixtures of different gases (e.g. carbon dioxide plus nitrogen). The choice of gas or gases will depend on the product characteristics and the types of microorganisms that require controlling. Expert advice should be sought to make sure the appropriate choice is made.

Two methods are commonly used to introduce the gas or gases into the product packaging; these are:

- flushing – a continuous stream of the gas mix is delivered into the packet, the air that was surrounding the food is pushed out, then the packet is sealed
- pre-vacuum technique – the air surrounding the food is removed from the packet (as per vacuum packaging), then the packet is refilled with the gas or mix before sealing.

Examples of products that may use MAP are fresh pasta products, fresh meat and seafood, and cook chill foods. Generally, plastic pouches or rigid trays that are sealed with film are used. Specialised equipment is required.

As with vacuum packaging, MAP is used mainly to maintain the quality of food; it is not generally applied to control the growth of pathogens. Therefore it is important to apply specific food safety controls to low-acid (above pH 4.6) food products (Box 69).

Active packaging

The term active packaging is used to describe packaging materials that are used to perform a role in addition to physical containment or protection. To further enhance the effectiveness of vacuum packaging or MAP using oxygen-free gas mixes, active packaging can 'mop-up' or 'scavenge' any leftover oxygen in the packet.

Box 69 – Chilled foods and controlling hazards associated with vacuum or modified-atmosphere packaging

One of the pathogens of concern for potentially hazardous food, *Clostridium botulinum*, can grow in vacuum packaged foods or under modified atmospheres (when there is no oxygen added). In fact, these packaging methods create almost an ideal environment for the growth of this pathogen. This is because *C. botulinum* grows in the absence of oxygen, and the growth of other microorganisms that would normally 'compete' with the *C. botulinum* is restricted.

Chilled vacuum or MAP foods that are low acid (above pH 4.6) must have specific control factors to prevent the growth of and toxin production by *C. botulinum*. These are in addition to maintaining products at 5°C or below.

If one or more of these control factors is not used, the shelf-life of these products must be restricted to 10 days or less at 5°C or below.

Specific control factors:

- Heat process of 90°C for 10 minutes or equivalent, achieved at the slowest heating point; either heated in-pack or followed by aseptic filling. See 'Packaging and heat processed products' on the following page.
- pH 5 or less throughout all components of the food (note – this will be difficult to achieve for products containing meat, fat or oil).
- A minimum salt level of 3.5% (in the aqueous phase) throughout all components of the food (Box 70).
- A water activity of 0.97 or less throughout all components of the food.

Other combinations of heat processing and recipe hurdles can be used if it can be proven by technical experts to prevent growth and toxin production by *C. botulinum*. Challenge testing and/or use of predictive modelling may be required. See under 'Determining appropriate use-by dates: technical experts' later in this chapter for more information.

This guidance was adapted from the United Kingdom's Food Standards Agency guidance on the safety and shelf-life of vacuum and modified-atmosphere packed chilled foods with respect to non-proteolytic *Clostridium botulinum* (see page 272 for details).

Box 70 – Determining the concentration of salt in the aqueous phase

The percentage of salt (i.e. sodium chloride, chemical formula NaCl) in the aqueous phase of a product can be determined using the following calculation:

$$\frac{\text{NaCl content} \times 100}{\text{NaCl content} + \text{moisture content}}$$

Definitions:

- NaCl content = grams of NaCl / 100 grams of product
- Moisture content = grams of water / 100 grams of product.

If you are not confident that you can perform this calculation accurately, you should seek advice from a technical expert.

Oxygen-scavenging active packaging has been found to be useful in foods where simple gas flushing cannot reduce the oxygen content sufficiently. This technology is used for porous foods such as long shelf-life breads.

Oxygen-scavenging active packaging can be in the form of specialised packaging films or sachets added to the packet before sealing. The sachets and film contain substances that react with, and absorb, oxygen.

If you feel active packaging may be suitable for your products, discuss this with your packaging supplier.

Packaging and heat processed products

Heat processing used in food manufacturing can be broken into two categories: those that achieve commercial sterility and milder pasteurisation processes. These have been discussed previously in Chapter 6. Details for packaging requirements for commercial sterility processes are not provided here because you must get product specific information from a technical expert.

Heating in-pack

Metal, glass or plastic packaging materials can be used when packaging foods before heat processing. Essential packaging requirements are the ability to be sealed hermetically, and for both the seals and packaging material to be resistant to the temperature used and to immersion in water.

Care has to be taken to leave an appropriate amount of headspace when filling. Guidance for headspace requirements for glass containers was provided in Box 66. The amount of headspace required for plastic packaging depends on the rigidity of the material used. For example, flexible pouches generally require the majority of the air in the headspace to be removed to reduce the chance of the pouch rupturing during heating and to improve heat penetration into the product. Packaging suppliers, or an appropriate expert, should be able to provide specific information about headspace requirements for different packaging types and product–process combinations.

When filling packs, it is essential that the same product weight or volume is consistently used. This must match the weight or volume used when initially determining your heat processing parameters. This ensures that the product receives the correct heat process each time. Checks can be performed on the filled weight or volume with the same packs used for checking seal integrity (see 'Seal integrity' later in this chapter).

Aseptic filling

The term 'aseptic' means free from contamination with microorganisms (i.e. sterile). Accordingly, correctly applied aseptic filling provides a high level of protection against post-process contamination.

Directly following heat processing, the product is filled into pre-sterilised containers in a sterile environment. This process requires specialised equipment and technical expertise.

Generally, aseptic filling is used with pumpable products such as liquids that are heat processed to achieve commercial sterility, cooled and then packaged. Some shelf-stable UHT products are produced with this heating–packaging combination.

Because pasteurisation does not achieve a sterile product, it is not commonly used in combination with aseptic filling.

Hot filling

If the product temperature following pasteurisation is hot enough during the filling process, the heat will kill the majority of any spoilage microorganisms that may contaminate the product after processing. The minimum filling temperature to achieve this is 85°C.

Hot filling prevents post-process contamination more effectively when used for high-acid (below pH 4.6) products, such as orange juice. If you intend to use hot filling for low-acid (above pH 4.6) potentially hazardous foods, it is strongly recommended you seek expert advice.

Products that are hot filled into containers with screw cap closures should be turned upside down for 3 minutes after capping so the heat from the product can kill microorganisms on the inner surface of the cap.

Some types of rigid plastic bottles and jars (e.g. PET) cannot be used for hot filling because the heat distorts them. This will be part of the decision-making process you and your packaging supplier undertake when choosing the packaging material suitable for your needs.

Packaging seals and closures

The type and form of packaging material used will restrict the type of seal or closure that can be used. The suitability of different seals and closures will also depend on whether products are heat processed in-pack or not.

Plastic pouch sealing

Plastic pouches or sachets are sealed by welding the packaging film. A combination of heat and pressure is used to weld the open ends of the pouch together. There are many different styles of equipment that can be used, such as a flat jaw sealer where the pouch opening is sandwiched between two 'jaws'. The jaws are fitted with heated strips and the combination of pressure, temperature and time then act to melt the plastic so it fuses together, sealing the package.

Preventing food particles from contaminating the seal area is an important step in protecting the seal integrity of plastic pouches. Food within the seal can act as a wick, drawing in moisture and microbial contaminants (see illustration on page 210). The presence of certain food substances in the seal area can also reduce the strength of the seal. If pouches are filled manually, it is very difficult to avoid food particle contamination of the seal area all of the time. It is important that packs are visually checked as they are filled so that any contamination can be detected. If pouches are filled and sealed automatically, adjustments can be made to ensure the equipment is operating effectively (e.g. product is not dripping from dispensing nozzles onto the seal area). If seal areas become heavily contaminated with food, the pouch should be discarded because it is not possible to effectively seal these packs.

If you find you cannot prevent food contamination around the sealing area, speak to your packaging supplier. They may recommend you change to a different type of plastic packaging that can still seal effectively despite this issue. Specialised packaging is available that can still be sealed even when product contamination would normally prevent proper sealing.

Food acting as a wick

Vacuum capping

Vacuum capping involves flushing the headspace of rigid containers with steam immediately before applying the lid or cap. When the steam condenses, the headspace pressure is reduced enabling a hermetic seal. The external air pressure pushes the lid firmly onto the container.

Vacuum capping can be done using equipment that injects a shot of steam into the headspace before heating in-pack, or as part of the hot-filling process (i.e. the steam is from the product itself).

The metal lids or caps of vacuum capped containers usually have a tamper evident 'pop-up button'. When the vacuum is formed correctly, the button is depressed. If the vacuum was poorly formed or is released by opening the container, the button is visibly raised. This system can be used by consumers when purchasing products as a way of checking the seal of the container has not been tampered with.

Standard screw caps (non vacuum)

Standard screw cap seals that are applied without a vacuum are generally used for products that do not require heating in-pack or hot filling, such as peanut butter or spice mixes. The lids or

caps can be made of metal or plastic. There are several tamper-indicating systems that can be used: common choices are shrink wrapping over the top of the cap or breakaway bands.

Seal integrity

Glass containers can't be commercially sealed without the aid of materials such as steel lids with polymer-based sealants. It is therefore important that you use the container-and-lid combination supplied; you should not swap to another type of lid unless you confirm it is suitable for use with glass. Lids for glass containers must never be re-used because the lid sealant may become worn or damaged during use or cleaning.

Heat processed packs must maintain the integrity of their seals to prevent post-process contamination. If you process products using a conventional canning process, seals must be inspected by a technical expert (such as an Approved Person, see under 'Canning' in Chapter 6) or by staff who have received accredited training.

Seals on products that have been pasteurised may be tested less rigorously, but it is still important that post-process contamination is avoided.

Specialised equipment can test seal integrity on-line. However, a variety of methods that don't need equipment can be used to check the strength and integrity of seals on flexible pouches, including:

- visual inspection – check that heat seals are free of gaps or creases, and food contamination
- pressure test – lay the pouch on a flat surface and press down on it firmly with open hand; check for leaks around heat seals
- immersion test – fully immerse pouch in a container of water, squeeze firmly with hands and watch closely to see if any air bubbles escape from the seals
- peel test – cut open the pouch and the remove the contents. A segment is then cut out, which is big enough to hold onto by the fingertips, with about a 1–2 cm wide portion of seal. Then attempt to pull apart the seal manually.

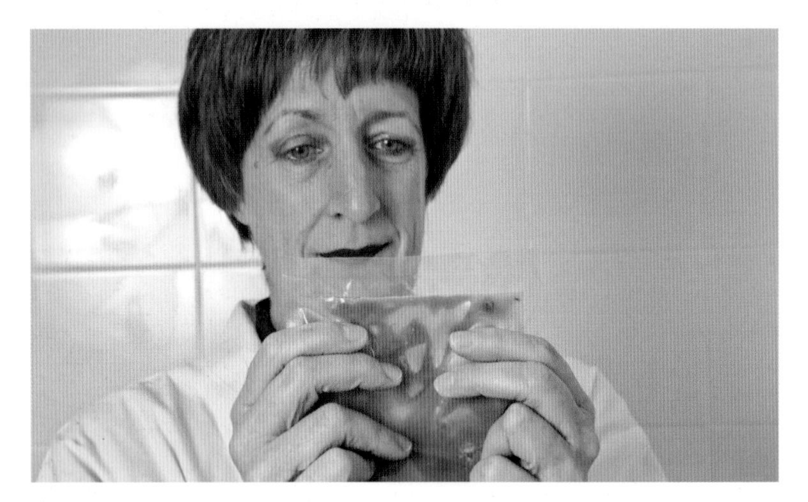

Visual inspection

Pressure test

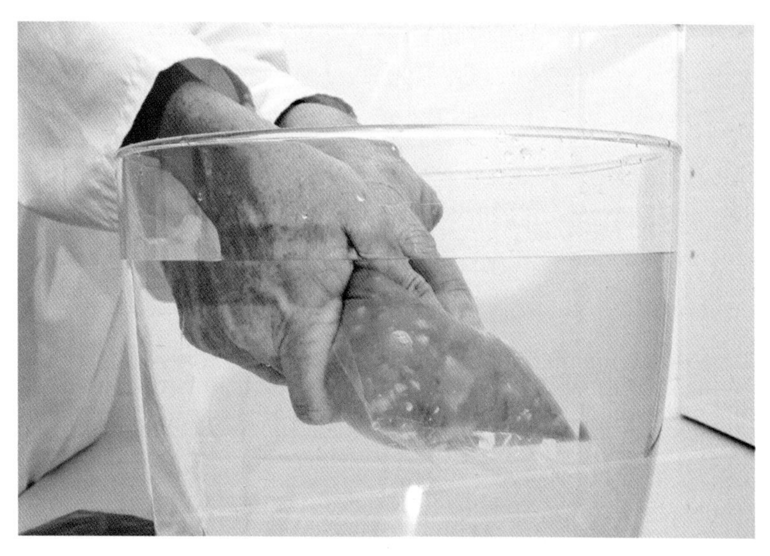

Immersion test

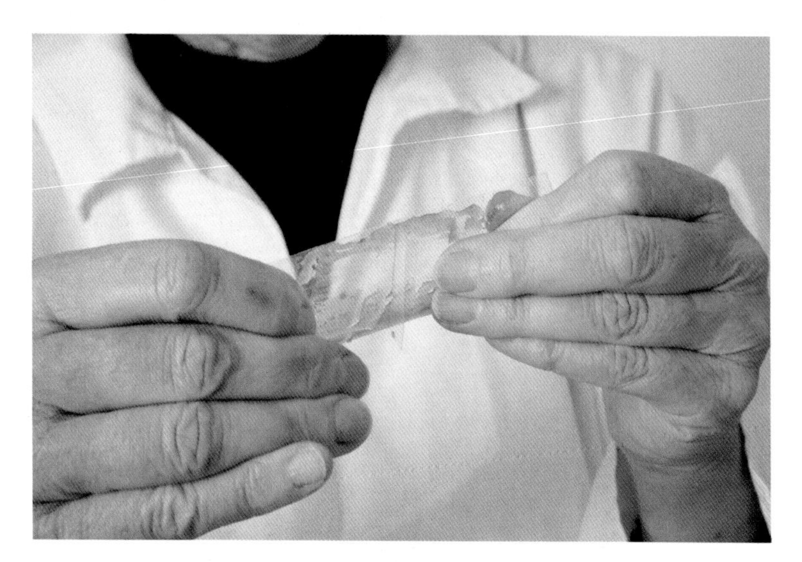

Peel test

The first few packages in every batch should have their seals inspected, and then periodic checks should be undertaken throughout the batch. Guidelines on the testing regime, including the number of packs to test per batch, should be provided by your packaging supplier or other technical expert. If you experience any problems with your packaging equipment or material, you should increase both the number and frequency of packages seal tested, until the issue is resolved.

Seal checks should be documented; an example of a form that could be used is shown in Box 71 but this is a simplified version for illustrative purposes only and does not incorporate full details (e.g. checking and recording product weight).

Secondary and tertiary packaging

The packaging described so far is termed primary packaging, which is packaging that is in direct contact with food. There are two other forms of packaging: secondary and tertiary.

Secondary packaging is optional, and includes packaging such as cardboard sleeves that fit over plastic tubs. The purpose of this packaging is to provide a surface for labelling and to protect the lid of the tub from being dislodged during handling.

Tertiary packaging is required to store and distribute products in bulk. Examples include cardboard boxes or pallets (or both may be used). This not only allows for your products to be delivered in bulk, but can offer additional physical protection to the products while they are being transported.

Requirements for tertiary packaging will in part depend on the strength and design of the primary and secondary packaging. For example, you would have to take into account any sharp edges that could create a hole in a packet pressing against another packet or how many glass jars can be stacked on top of each other without breaking or deforming lids.

Using contract packers

Contract packers or co-packers, who process and package foods, are often used by small businesses that lack the equipment or expertise to process or package their products independently. This is particularly the case for those just starting off their businesses whose knowledge about how to process their product to give it the required level of safety and quality may be limited. Products that fall into this specialist need category are those that require UHT equipment and aseptic filling, and those that require a conventional canning process. Using the services of a co-packer will also assist small businesses to meet the requirements of their suppliers, such as provision of auditable HACCP plans.

Some major food manufacturers provide co-packing services, as long as the product does not compete with their own labels. Dedicated co-packers are also available.

Box 71 – Example seal integrity inspection record sheet

SMITH & SONS

SEAL CHECKS RECORDING SHEET FOR RETORTED STEWS

Method:

Visual check – if misalignment, creases, gaps or food contamination are visible, contact supervisor immediately for corrective action instructions. DO NOT recommence sealing until you receive their approval.

Pressure test – if leaks occur, notify supervisor immediately for corrective action instructions. DO NOT recommence sealing until you receive their approval.

Immersion check – if bubbles are visible, notify supervisor immediately for corrective action instructions. DO NOT recommence sealing until you receive their approval.

Number of packs to test and frequency:

Standard – test 2 packs at the start of each product batch, then every 1 in 50 per run

Additional – test 5 extra packs after equipment maintenance or adjustment of settings

Pack number	Visual	Pressure	Immerse	Corrective action	Staff initial	Supervisor initial
Date: *30/11/2010*		Product: *Chicken stew*		Retort No: *2*	Batch: *1298-7*	
1	*Pass*	*Pass*	*Pass*		*MW*	
2	*Pass*	*Pass*	*Pass*		*MW*	
50	*Pass*	*Pass*	*Bubbles – fail*	*KT peel test – fail Adjust heat settings on sealer from 5 to 6*	*MW*	*KT*
51	*Pass*	*Pass*	*Pass*	*KT peel test – pass*	*MW*	*KT*
52	*Pass*	*Pass*	*Pass*	*KT peel test – pass*	*MW*	*KT*
53	*Pass*	*Pass*	*Pass*		*MW*	*KT*
54	*Pass*	*Pass*	*Pass*		*MW*	*KT*
55	*Pass*	*Pass*	*Pass*	*KT – approval to continue production*	*MW*	*KT*
100	*Pass*	*Pass*	*Pass*		*MW*	

Remember, you are legally responsible for the safety of products you sell. You will need to establish a detailed legal agreement with the co-packing business before using their services. This should include specifications covering issues such as compliance with the Code, raw ingredient quality, processing parameters, temperature control and control of hazards (e.g. food allergen cross-contact contamination).

Product shelf-life and food safety

One of the most difficult tasks for a food business is determining the appropriate shelf-life for each of their products.

The term shelf-life refers to how long a food can be stored under specified storage conditions without becoming a food safety hazard or developing unacceptable quality attributes. Retailers, consumers and other users of food products can tell how long a product's shelf-life is by looking at use-by or best-before dates on packaging or labels. These dates are nominated by the manufacturer and are generically called 'date marks'.

Products that have a shelf-life greater than 2 years are exempt from needing a date mark, in addition to some other product types listed in the Code (Standard 1.2.1 Application of Labelling and Other Information Requirements). Products that require date marking need to be labelled with a use-by date, best-before date or baked-on date (bread only).

Box 72: Definitions of use-by and best-before date

Use-by date: in relation to a package of food, means the date which signifies the end of the estimated period if stored in accordance with any stated storage conditions, after which the intact package of food should not be eaten because of health and safety reasons.

Best-before date: in relation to a package of food, means the date which signifies the end of the period during which the intact package of food, if stored in accordance with any stated storage conditions, will remain fully marketable and will retain any specific qualities for which express or implied claims have been made.

Source: Standard 1.2.5 Date Marking of Food

A use-by date is the date mark required if the shelf-life is specified for food safety reasons. If the shelf-life is needed for quality reasons, a best-before date is required. See Box 72 for the definitions of use-by and best-before date provided in the Code. The focus here will be on how to determine a safe shelf-life, and therefore use-by dates, for your products.

Do your products require a use-by or best-before date?

If there is a possibility that pathogenic bacteria could be present in your final product – and these could grow or produce toxin to unacceptable levels in the product at the intended storage temperature – then the Code requires that you label your product with a use-by date (Standard 1.2.5). Remember, viruses can't grow in food, which is why only bacteria are referred to here.

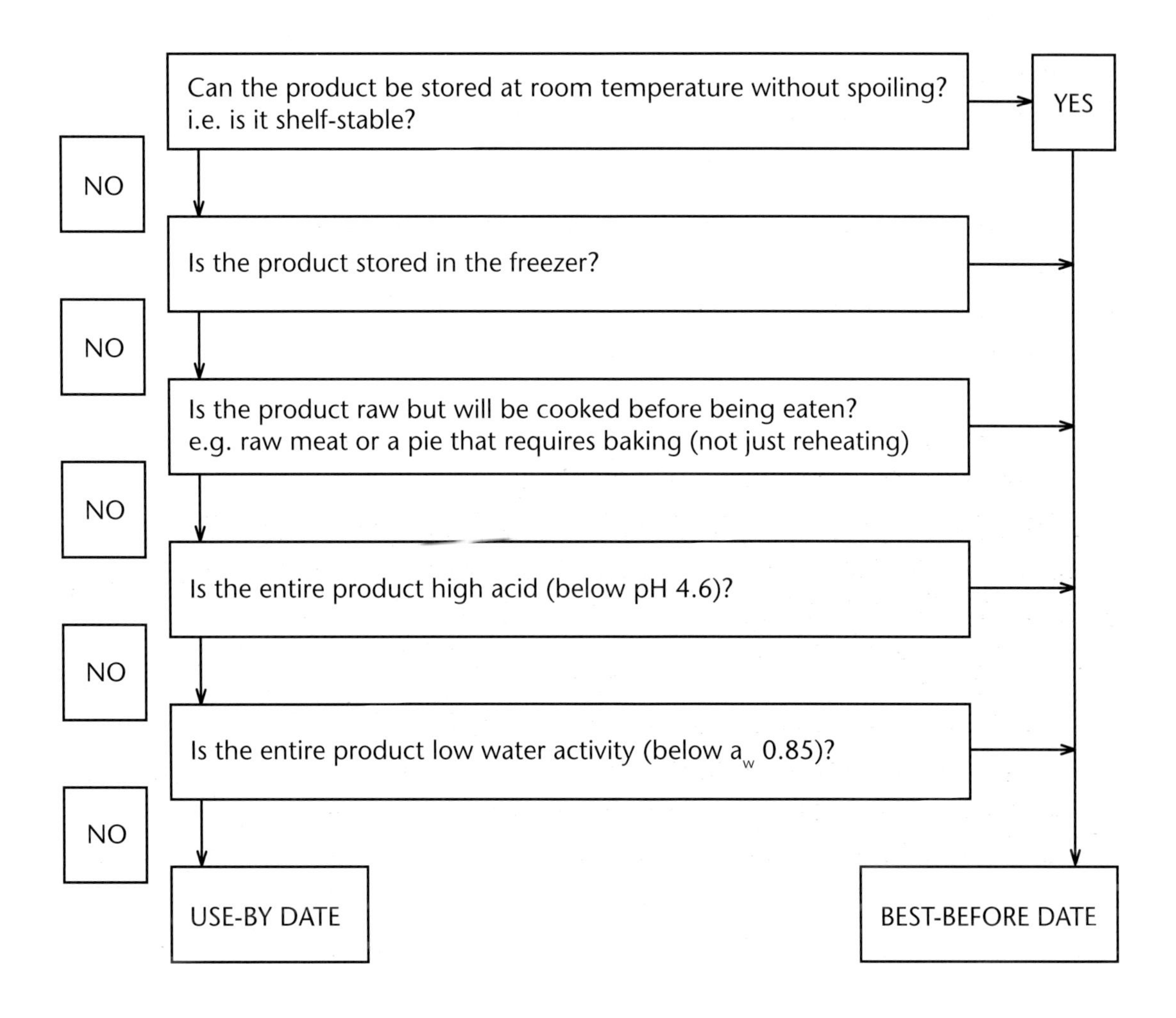

It is the responsibility of a food business to determine if its products require use-by dates and, if so, what the dates should be. These are legal requirements specified in the Code (Standard 1.2.5). State and territory authorities regulate their implementation. You may be requested to supply evidence on how you made these decisions: either by an EHO or by the retailers who stock your products or you wish to have stock your products.

The decision tree on page 217 will help you work out if use-by or best-before dates are required for your products. It is based on the assumption that chilled or refrigerated ready-to-eat foods may contain pathogens of concern (see Box 73).

If, after comparing your products to the decision tree, you are still not sure if a use-by or best-before date is required, you should seek the assistance of a technical expert, such as a food microbiologist. Additional guidance is also provided in the FSANZ document, User Guide to Standard 1.2.5 – Date Marking of Packaged Foods (see page 270 for details).

In general, most potentially hazardous foods require a use-by date, including ready-to-eat cook chill foods, because they:

- are often low acid (above pH 4.6), or contain a mixture of low-acid and high-acid foods (e.g. meat in a tomato-based sauce),
- have high water activity and/or
- are either not heated at all, or only re-heated by consumers.

Box 73 – Pathogens of concern (for refrigerated foods)

As discussed earlier, there are some pathogenic bacteria that can grow slowly under refrigeration. These are *Listeria monocytogenes*, and some types of *Clostridium botulinum* and *Bacillus cereus*. Because it is very difficult for any food manufacturer to be absolutely certain these could never be present in their products, it is best to err on the side of caution and assume that they may be.

Due to the unfavourable conditions, these pathogens grow and/or produce toxin very slowly when stored at 5°C or below. Therefore, there is some time before they reach high enough levels to cause illness and, in some cases, the food will spoil first. However, this is dependent on many factors and, again, it is best to err on the side of caution and apply a use-by date if this is what is indicated by the decision tree.

You may choose to consult a technical expert, such as a food technologist or food microbiologist, who can provide further guidance based on your specific products.

Determining appropriate use-by dates: in house

If you need to apply a use-by date to one or more of your products, you are responsible for determining what the safe shelf-life is. Use-by dates specify the date beyond which the product may no longer be safe to eat.

The time period starts from the day of production. For example, if you determine that it is safe for your product to have a 10 day shelf-life and it takes one day to prepare and package and one day to transport, then it would have 8 days shelf-life remaining at the retail outlet or in the consumer's home. It is illegal to display or sell products after their use-by date has passed.

The safe shelf-life of a food is influenced by a number of factors including:

- the microorganisms present in the raw ingredients (type and level)
- the product recipe; specifically pH, water activity, preservatives, chemical composition and physical structure
- how the product has been processed (e.g. if a heat process or other kill-step was applied)
- how hygienically the product was prepared, processed and packaged

- the packaging method and type of material used
- the availability of oxygen or the presence of other gases that inhibit microbial growth
- the temperature used for transport and storage of the final product, including possible temperature abuse after it leaves the manufacturer
- the stability of nutrients; however, this only relates to the use-by date of products whose nutritional profile is critical to the health of a consumer, such as specialised products used as the only source of food for hospital patients.

Pasteurised products

The packaging method used in conjunction with pasteurisation influences the safe shelf-life of products. Heating in-pack and aseptic filling significantly reduce the chance of post-process contamination with pathogenic bacteria. If the product temperature is above 85°C during hot filling, then the chance of post-process contamination is greatly reduced. However, there is still a chance of post-process contamination during hot filling, especially with heat resistant spores, and this must be considered when you determine the shelf-life.

Guidance on shelf-life determinations for products that are processed using specific time–temperature combinations and packaging techniques has already been provided (Chapter 6, under 'Cook chill'). If your products do not fall into these categories, you will need to seek expert guidance to help you determine correct use-by dates; see 'Determining appropriate use-by dates: technical experts' below.

Vacuum or modified-atmosphere packaged products

Potentially hazardous foods that are vacuum or modified-atmosphere packed must have their shelf-life restricted to up to 10 days at 5°C or below, unless:

- they have received a 90°C for 10 minute or equivalent heat process, either in-pack or followed by aseptic filling
- other appropriate control factors specified in Box 69 (e.g. pH, salt and/or water activity) are used; most cured meats, such as ham, would fall into this category.

Determining appropriate use-by dates: technical experts

In general, the technical experts who perform analyses to establish appropriate use-by dates are food microbiologists. Microbiologists can use two different methods: predictive modelling and/or challenge testing.

The questions the microbiologist needs to answer are:

- Which pathogens (if any) are likely to be present in the product?
- At what level are they likely to be present?
- Are these pathogens able to grow in the product (at specified storage temperatures)? If this answer is 'Yes',
- How long may it take these pathogens to grow to levels that are likely to cause illness?

It may be necessary for the laboratory to do experiments to gather this information as described on the next page. Once these questions have been answered to their satisfaction, the experts will be able to provide written recommendations for the safe shelf-life of your products. These can be used to show an EHO, if they request information about how you determined your use-by date.

If you change your recipe proportions, ingredients or processing method you should again seek expert advice to determine if you need to re-examine the appropriateness of your current use-by dates.

Predictive models

Microbiologists have developed microbial predictive models that can assist in shelf-life determinations. To develop the models, data from laboratory studies testing the ability of pathogenic microorganisms to survive or grow under different conditions (e.g. varying pH, water activity, temperature or packaging atmospheres) is collected. The data is then analysed using mathematics and statistics. Complex mathematical equations, called models, are then produced. These can be used to predict the likelihood of specific pathogens growing in food products, and at what rate under specified storage conditions, hence the term 'predictive models'.

Some predictive models have been developed into computer software, which can be purchased or downloaded free of charge. However, small business operators should not use these models without expert assistance. Even with expert assistance, it is generally not satisfactory to base shelf-life determination solely on model predictions. Although they are valuable tools, these models do have limitations. These are partially due to 'unknowns' in food products and processing conditions, such as the types or levels of pathogens present in raw ingredients, or the effects of antimicrobial compounds that may be present in food naturally.

Models may still be used to guide decisions made during the product development phase. This can then help reduce costs associated with 'real life' laboratory testing. For example, there may be several different pH and water activity combinations able to enhance your product's quality.

Instead of testing all of these to find the combination that will deliver the safest product, predictive modelling could narrow down the choices.

Challenge testing

Challenge testing needs to be done by a laboratory with specific expertise and involves:

- deliberately contaminating product samples with a specific level of the pathogens of concern
- storing the product under controlled conditions (e.g. temperature and atmosphere)
- testing to monitor the changes in the levels of microorganisms, whether it is growth or death.

One of the first questions you will be asked by the laboratory you choose to challenge test your products is 'What is your desired shelf-life?' It will be useful for you to consider the following to help guide your decision:

- The laboratory performing the analysis can provide you with guidance on what a reasonable shelf-life may be for your product type.
- If your product normally spoils after about a month's storage, there is no point in trying for a 6 week shelf-life.
- You can look at the use-by date on other products that are similar to your own.
- The retail outlet that you plan to sell your products to may specify a minimum shelf-life.

You will also need to provide the laboratory with information about your products, including:

- the types of ingredients used, including preservatives and other antimicrobial agents
- the values for the pH or water activity (if not known, the laboratory will need samples to test)
- the details of any heat process used
- information about the packaging materials and method
- whether you have had any food safety issues with your products, or pathogens detected at your premises, in the past.

The information you provide will then be assessed by the laboratory to determine the pathogen types to use in the challenge test. They may also review published data, including any information on outbreaks of foodborne illness associated with similar types of products and on the types of pathogens likely to be present in the raw ingredients you use.

The laboratory may ask you to make special batches of your products to use for the challenge tests. This is because they may recommend testing products that represent the 'worst case scenario' for your product recipes. For example, if the pH of your products normally ranges from pH 5 to 5.5, they may specifically request product samples that are pH 5.5 because the pathogens will be more likely to grow and/or produce toxin at this pH. Similarly, the temperature at which the laboratory chooses to store the products during the testing phase may include a 'worst case scenario', such as using 8°C for products that should be stored at 5°C or below.

Once all of the pieces of information are put together and you have provided the product samples, the laboratory will be able to commence the testing. After inoculating the product with the pathogen(s), they will store samples at the specified temperature(s). They will then regularly test samples to see if the pathogens have grown. If no growth is detected, they will keep on testing the samples until the end of, or beyond, the desired shelf-life. If pathogen growth is detected before the end of your desired shelf-life then the product and shelf-life combination is a potential food safety issue. You must reduce your shelf-life and/or change your product, depending on the advice the laboratory provides you.

Sometimes other microorganisms already present in the product samples may interfere with the test by growing to high levels, therefore possibly affecting the ability of the testing laboratory to monitor pathogen levels. These other microorganisms are usually spoilage microorganisms and growth to high levels during a challenge study can indicate that the desired shelf-life is too long and you may need to rethink your recipe or processing method. However, a benefit of having spoilage microorganisms growing in the product is that they can compete with, and outgrow, pathogens, thus adding to the safety of the product.

Why not use shelf-life testing to determine use-by dates?

During shelf-life testing, also called storage trials, product samples are stored and tested for specific changes over time. Accelerated shelf-life testing is a special form of shelf-life testing that can provide an indication of a product's shelf-life in a shorter time period. Growth of any microorganisms that were present in the samples at the end of processing and packaging is usually monitored, in addition to changes in the products quality attributes.

However, shelf-life testing is most useful in studying changes to product quality for determining best-before dates. It is generally not recommended for use as a tool to determine use-by dates. Storage trials will only monitor pathogen growth if the pathogens were present in the particular batch of product being tested. Information on the growth of other pathogens that may be present in your product from time to time will not be captured. Product samples used for shelf-life testing are chosen during a normal daily production run, so they are representative of a typical sample. Depending on the raw ingredients used, and if you strictly follow hygienic practices, there may be minimal chance of your products being contaminated with pathogens. However, the chance of pathogens being present may never be completely eliminated; otherwise the products would not require a use-by date! As such, you need to determine what the potential is for pathogens of concern to grow in your product. This evidence can be provided through challenge testing, discussed earlier.

Labelling your products

The Code specifies the legal requirements for food labels; only those that are specified for food safety reasons are discussed here. As the information provided here is only a limited guide, you should seek independent advice from an EHO, technical expert or legal adviser on whether your labels comply with all requirements. If you are planning to export your products, requirements of the country you are exporting to need to be considered. AQIS can provide details of the regulatory requirements for other countries.

It is essential to provide the correct information on product labels, so that:

- retailers and consumers know the correct temperature to store the product, to reduce the risk of pathogens growing to unacceptable levels if the storage temperature is too high
- retailers and consumers know when the product must be discarded (i.e. when the use-by date is reached)
- consumers who are allergic to food allergens can be alerted to their presence
- products can be quickly traced in the event of a recall
- consumers know how to safely prepare and handle the product; this is particularly important for products that require cooking in the home.

Labels can be applied to primary, secondary or tertiary packaging. You should seek guidance from the above listed experts on what the minimum labelling requirements are for each layer of packaging used and your specific product types.

Mandatory statements and declarations

Products containing certain substances, as listed in the Code, must be labelled with specific information. If you are uncertain if these apply to your products, you must seek advice from an EHO, your ingredient supplier or a technical expert.

Food or ingredients that may require specific label information are:

- food allergens (see 'Food allergens' below)
- bee pollen

- cereal-based milk and beverages (intended use as cows' milk replacement)
- evaporated and dried products made from soy or cereals (intended use as cows' milk replacement)
- aspartame or aspartame-acesulphame salt
- quinine
- caffeine
- guarana or extracts of guarana
- phytosterol esters or tall oil phytosterols
- propolis
- unpasteurised milk, liquid milk products and egg products
- royal jelly
- polyols or polydextrose.

Source: Standard 1.2.3 Mandatory Warning and Advisory Statements and Declarations

For example, the milk labelled with the information below does not contain a high enough fat content to meet the complete nutritional requirements of infants less than 2 years old, and therefore an advisory statement is required.

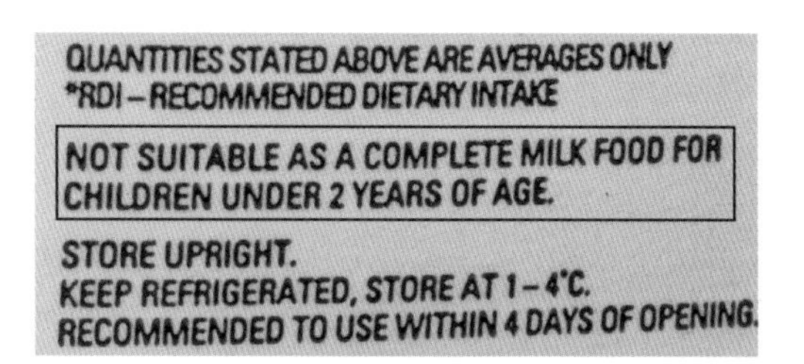

Food allergens

Because food allergens are commonly used in many different products, detailed guidance on how to correctly label products containing them will be provided here. The information provided in this section has been sourced from the Code and the Australian Food and Grocery Council (page 277). You should seek advice on whether your labels comply with the Code requirements from an EHO, technical expert or legal adviser.

The presence of certain substances must be declared on the label of packaged food. These are:

- cereals containing gluten and their products, namely wheat, rye, barley, oats and spelt
- crustaceans and their products
- egg and egg products

- fish and fish products
- milk and milk products
- peanuts and soybeans and their products
- added sulphite, if more than 10 mg/kg of food present
- tree nuts and sesame seeds and their products.

Source: Standard 1.2.3 and 1.2.4 Labelling of Ingredients

Allergens can be present in products in the form of an ingredient, additive or processing aid. Their presence must be declared on the product label regardless of how little is added to the product.

To meet the requirement of 'declaring the presence' of these substances, the Code specifies that they are listed in the ingredients list. However, it is strongly recommended, and it is common practice, to list them separately on the label to further warn consumers of their presence. Box 74 shows an example of a recommended label format, including the labelling recommended if a risk-assessment process, such as the VITAL system, shows there is sufficient evidence that an allergen may be present through cross contact.

Fully descriptive declarations are required for some ingredients so that individuals who are affected by food allergens can be completely informed. Specifically this applies to:

- cereals – the specific name of the cereal (e.g. 'wheat')
- vegetable oils – if the source of the oil is peanut, soybean or sesame, the specific source name
- crustaceans – the specific name of the crustacean (e.g. 'prawns')
- nuts – the specific type of nut (e.g. 'peanut')
- starch – if the source is from any of the previously listed cereals then the specific name of the source cereal.

If you use ingredients that are themselves made up of two or more ingredients, it is a requirement of the Code that you list these ingredients separately. This ensures that the presence of food allergens is not 'hidden'. For example, for canned spaghetti, you would need to separately list the ingredients of the spaghetti, 'spaghetti (wheat flour, egg, water), meat, sugar, water' so consumers know there is wheat flour and egg present in the product.

Once you have finished with a roll of labels at the end of a production run, it is important they are returned to their storage area. If they are left in the labelling machine, there is a danger they may accidentally be used for the next product, which may be allergen-free or may contain allergens different from those stated on the label.

Box 74 – Recommended labelling format for food allergens

INGREDIENTS

Water, vegetable oil, vinegar, cane sugar, tomato paste (5%), salt, parmesan **cheese** (2%), **egg** yolk, maize thickener (1412), **almonds**, red capsicum, **soybean** oil, garlic (1.0%), vegetable gum (415), spice, herbs, **wheat** cornflour, flavour (**wheat** maltodextrin, **sesame** oil), antioxidant (320).

Contains milk, egg, almonds, soy, wheat and sesame.

May be present: xxx. *(only used if identified by VITAL to be necessary)*

Source: Food Industry Guide to Allergen Management and Labelling (see page 271)

Also, if you alter a label because you have changed a recipe, it is particularly important that the old labels are discarded if the new recipe includes a food allergen that was not present previously.

Boxes for different types of packaging or labels should be clearly labelled so they can be identified easily.

Baby foods

There are specific labelling requirements for infant foods and infant formulas. These requirements are specified in the Code: Standard 2.9.1 Infant Formula Products and 2.9.2 Foods for Infants. The requirements for infant foods only will be provided here; you must seek expert advice if you wish to produce infant formula. An 'infant' is a child less than 1 year old.

Requirements specific to infant food are that the labels:

- must not include any reference (stated or implied) to it being suitable for infants less than 4 months old
- must indicate the consistency of the food (e.g. 'pureed')
- must state the minimum aged infant the food is recommended for in numbers (e.g. '4 months' instead of 'four months')
  - if this is infants between 4–6 months, the following words must be included: 'Not recommended for infants under the age of 4 months'

- must include the word 'sweetened' if the added sugar content is more than 4 g/100 g
- must include the word 'sterilised' in association with the word 'honey' (if it is an ingredient)
- must include the percentage of the food source of protein in the food if it is referred to on the label
  - food sources of protein are: milk, eggs, cheese, fish, meat (including poultry), nuts and legumes
- must include the words 'Not suitable for infants under the age of 6 months' if the food contains more than 3 g/100 kJ of protein
- must not claim (stated or implied) that the food is a source of protein unless more than 12% of the average energy content of the food is from protein
- must not represent the food as being the sole or principal source of nutrition for babies
- must not recommend that the food can be added to bottle feeds of infant formula
- must not compare (stated or implied) the vitamin or mineral content of the food with that of any other food
- must not make any claims about the vitamin or mineral content of the food unless the requirements of the Code are met (Standard 2.9.2)
- must set out the nutrition panel in the format specified in the Code (Standard 2.9.2)
- must include directions on how dehydrated or concentrated products are reconstituted
- must include storage instructions covering the period after it is opened.

Smallgoods

The Code (Standard 2.2.1 Meat and Meat Products) specifies labelling requirements for fermented smallgoods. These are intended to inform consumers when products have not been heat processed or cooked, because those who are at higher risk of contracting a foodborne illness may choose to avoid eating these.

The requirements are to use the following prescribed names, which have to be used exactly as worded here, for:

- Fermented comminuted processed meat (Box 75)
  - not heat processed or cooked: 'Fermented processed meat – not heat treated'
  - heat processed: 'Fermented processed meat – heat treated'
  - cooked: 'Fermented processed meat – cooked'.
- Fermented comminuted manufactured meat (Box 75)
  - not heat processed or cooked: 'Fermented manufactured meat – not heat treated'
  - heat processed: 'Fermented manufactured meat – heat treated'
  - cooked: 'Fermented manufactured meat – cooked'.

Box 75 – Processed and manufactured meat

Processed meat is a meat product containing at least 300 g/kg of meat, where the meat has undergone a method of processing other than boning, slicing, dicing, mincing or freezing.

Manufactured meat is processed meat containing at least 660 g/kg of meat.

Source: Standard 2.2.1

- Fermented comminuted manufactured meat that has a trade name; the trade name must be associated with the following words
  - not heat processed or cooked: 'Fermented'
  - heat processed: 'Fermented heat treated'
  - cooked: 'Fermented cooked'.

Product traceability

If food products are suspected of causing foodborne illness, or there is any reason to believe they may be hazardous if eaten, they may be recalled (see Chapter 8 for further information). Usually this will only be for a single product batch or for products manufactured over a specified time period. Warehouses, retail outlets and consumers that are holding these products must be able to identify the implicated items so they can discard them, or return them to the business or place of purchase.

Lot or batch identification

The Code specifies that labels on packaged foods must include lot identification (Standard 1.2.2 Food Identification Requirements).

Lot numbers, also called batch numbers, are unique codes used to identify:

- a single day's production run (for businesses that make one product type)
- different batches of the same product made on the same day
- different product types.

Use-by or best-before dates can be used instead of batch numbers, as long as only one batch of each specific product is made per day.

There are two exceptions to this requirement: individual portions of ice cream or ice confection, and products in small packages (less than 100 cm^2). However, there must be a lot number on the outer bulk packaging or container in which these small packages are contained during storage and retail display.

Name and address of supplier

The Code also specifies that labels on packaged food must include the name and the address in Australia or New Zealand of the business that supplies the food (Standard 1.2.2). This is the name and business address of the food business that manufacturers or packages the product (if made and packaged in Australia or New Zealand) or the name and business address of the company that imported the product (for products manufactured overseas).

This information will be used in conjunction with the lot number to identify products that are involved in a recall. It will also be used by authorities to notify a food business of any issues associated with their products.

Additionally, this information can be used by consumers to inform food businesses if they identify any safety issues with their products (e.g. presence of broken glass).

Directions for use and storage

The Code specifies that directions for use and/or directions for storage of food must be included on the label if required for reasons of health and safety (Standard 1.2.6 Directions for Use and Storage).

Product categories that the Code specifies must be labelled with information for consumers and retailers include those that require:

- a use-by date and storage temperature
- cooking instructions
- thawing before cooking or re-heating
- post-opening instructions (e.g. infant or baby foods).

Use-by date and storage temperature

If your product is required to be labelled with a use-by date, you must use the words 'use-by'. You cannot use any other term. These words must be followed by the date or instruct the consumer where to look on the product (e.g. 'See base').

The information provided and the form of the date must meet these minimum requirements:

- Use-by less than 3 months – the day and the month e.g. 'Use-by 3 Dec'.
- Use-by greater than 3 months – the month and the year e.g. 'Use-by Dec 2010'.
- The month may be written in letters, but the day and the year must be written in numbers e.g. 'Use-by third Dec' is not acceptable.
- The day, month and year must be clearly distinguishable; e.g. 'Use-by 12/12' is not acceptable because it may be interpreted as either use-by the 12th of December or by December 2012.

Apart from a manufacturer's or packer's code, no other date should appear on the label of packaged food.

All products with a use-by date are required by the Code to be labelled with the storage temperature required to achieve this shelf-life. In nearly all cases, this will be at 5°C or below. The choice of words is not specified, so you may wish to emphasise the importance of storing at this temperature by saying 'Keep refrigerated at 5°C or below at all times'.

Cooking and re-heating instructions

It is essential that you provide clear and simply written instructions on any products that require cooking or re-heating by consumers. This is particularly important for products falling into the potentially hazardous food category that are sold raw and require cooking by consumers as a

Box 76 – Determining heating instructions for microwavable products

- Trial samples should be representative of the final product – this includes recipe, weight range, component distribution and packaging.
- The temperature of the product should be stabilised, equivalent to a consumer's fridge or freezer, before testing. Test and record this temperature before commencing the trial. It is also preferable to use two different starting temperatures, incorporating a 'worst case scenario', which will be colder than the standard temperature.
- Use a stopwatch for timing; do not rely on the timer on the microwave because this may be inaccurate.
- Record the final product temperature at several points (including the slowest heating point), and full details of the method.
- Assess sensory qualities – if the product smells, looks or tastes too overdone, people are likely to heat less.
- Test at least in duplicate, in a minimum of four different ovens (perform tests in a cold microwave).
- Consider labelling with more than one different oven wattage – with a spread wide enough so that consumers can estimate the appropriate time for wattages not listed.
- Other variables to consider include:
  - microwave power levels and heating time: you should use different microwaves to trial different wattages
  - initial temperature of the product: if your trials are with frozen products, you must state in the instructions that the product must not be thawed before heating
  - stirring in middle or towards end of the heating time: the product must be thawed enough by this stage for adequate mixing
  - standing time: may be incorporated to achieve your desired time–temperature combination
  - positioning on the microwave plate: you will need to trial positions in the middle and the edge because these affect heating rates
  - consider if the lid needs to be removed or simply pierced before heating.

Source: adapted from Guidelines on the verification of re-heating instructions for microwaveable foods (see page 272)

food safety control step. However, it should be noted that this practice should generally be avoided. It is preferable for these products to be cooked on your premises and only re-heated by consumers, because you cannot guarantee that consumers will follow your instructions.

The Code specifies that products containing raw meat, fish or poultry that have been joined to look like a whole cut or fillet must be labelled with cooking instructions indicating how microbial safety of the product can be achieved (Standard 2.2.1 and 2.2.3 Fish and Fish Products). In most cases, this involves heating the food to achieve 70°C for 2 minutes (or equivalent) at the slowest heating point. Because these products actually look like whole pieces, there is a risk consumers will simply cook the outside until browned, not realising that thorough cooking throughout is required. Similarly, uncooked products coated in breadcrumbs must have clearly stated cooking instructions. If these products have been flash fried (to brown the crumbs) they may appear to be cooked even though they are still raw inside. In Australia in 1998, there was an outbreak of salmonellosis caused by undercooking of crumbed chicken nuggets by consumers: 18 people became ill.

Going to a supermarket and looking at heating instructions for similar type and size products to your own is a good starting point. However, you must use these as a guide only because these instructions may be incorrect or not suitable for your product. You must always do your own trials in a domestic oven or microwave so you can determine the correct method needed to reach the desired temperature at the slowest heating point. Box 76 shows some tips for determining heating instructions for products to be heated in a microwave.

Due to specific hazards associated with raw bamboo shoots or sweet cassava, the Code specifies that they must be labelled (or accompanied with) these directions:

- Bamboo shoots (raw) – should be fully cooked before being eaten.
- Sweet cassava (raw) – should be peeled and fully cooked before being eaten.

Source: Standard 1.2.6

Post-opening storage and shelf-life instructions

The FSANZ User Guide to Standard 1.2.5 – Date Marking of Packaged Foods (page 270) states that providing '… advice to consumers on the shelf-life and storage of a food after it is opened ... is compulsory where necessary for reasons of health and safety'.

Any products that are low acid (above pH 4.6) and high water activity (above a_w 0.85), should be labelled with specific post-opening storage and shelf-life instructions.

Although products such as canned foods have been processed to achieve commercial sterility, as soon as they are opened there is the potential for pathogens to contaminate and grow within the product.

Storage instructions should include a requirement to store at 5°C or below after opening. A restriction on the length of storage time, such as 'Use within 48 hrs of opening' should also be included. If you wish to instruct consumers to store for longer than this after opening, it is recommended that you seek expert guidance.

Legibility of writing on labels

It is important that the information provided on food labels is large and clear enough for the average consumer to read. This is particularly true for any information provided for food safety reasons. The Code (Standard 1.2.9 Legibility Requirements) specifies minimum requirements for label legibility:

- Written information must stand out clearly from the background.
- It must be written in English.
- When it is also written in another language, the information in the other language must not negate or contradict the information written in English.
- Warning statements (Box 77) must be written in:
  - normal size package – type that is at least 3 mm high
  - small package – type that is at least 1.5 mm high.

The types of issues that may prevent compliance with the Code requirements:

1. The ink used to print the use-by date has smeared.
2. The colour of the writing is similar to the background colour and there is too little contrast.
3. The writing is sitting too close together.
4. A logo is printed over words obscuring some of the words
5. A typeface is used that is difficult to read.

Here are some examples of different ways you can make important information stand out:

1. Bolding food allergen names in ingredients lists.
2. Using uppercase (e.g. MUST BE FULLY THAWED BEFORE COOKING).

Box 77 – Warning statements

A warning statement is information required on certain food product labels, as prescribed in the Code for:

- royal jelly (Standard 1.2.3)
- kava (Standard 2.6.3 Kava)
- infant formula (Standard 2.9.1)
- infant food (Standard 2.9.2)
- formulated supplementary sports foods (Standard 2.9.3 Formulated Meal Replacements and Formulated Supplementary Foods).

If you make any products that fall into these categories, it is your responsibility to make sure you review the appropriate section of the Code and label your products correctly.

3. Putting the information inside a box with a border.
4. Using a typeface colour that clearly stands out.

When you are designing your labels, you may wish to seek the opinion of friends and family to check that a wide range of people can easily read the information provided.

Minimum recommended labelling

Some packaged products for retail sale are not required by law to have a label (The Code, Standard 1.2.1). However, it is recommended that any potentially hazardous food be labelled with a certain minimum amount of information.

It is advisable for any potentially hazardous foods to be labelled with a use-by date and the statement 'Keep refrigerated at or below 5°C'. This is specifically recommended for products that are:

- made and packaged on the premises where they are sold
- cut fresh fruit and vegetables
- sold at fund-raising events.

It is also recommended that any products that contain food allergens have the name of the allergen clearly stated on the label (Box 74).

Requirements for unlabelled products

The two product categories specified in the Code as exempt from having to be labelled are:

- unpackaged foods
- packaged foods, but the surface is too small (i.e. less than 100 cm^2).

These products must have certain information either displayed in connection with the food, provided to the purchaser (whether requested or not) or provided to the purchaser upon request.

The information provided here is intended to alert you to these requirements; detailed instructions on how to implement them are not provided. Therefore, if your products fall into any of these categories, you should seek specific advice from your state or territory food authority or an EHO.

A summary of the relevant requirements for unlabelled products (that relate to food safety are):

- Warning and advisory statements or declarations (Standard 1.2.3) – such as the presence of food allergens, presence of other substances that may cause allergic reactions (e.g. bee pollen, propolis or royal jelly) and the presence of unpasteurised egg or egg products.
- Directions for use or storage (Standard 1.2.6) – if needed for reasons of health and safety.
- Cooking instructions for raw meat (including poultry) and fish that has been formed or joined to look like a whole cut or fillet of meat or fish (Standards 2.2.1 and 2.2.3).
- The prescribed name for comminuted fermented meats must be displayed (Standard 2.2.1).

KEY MESSAGES FROM CHAPTER 7

- Poorly sealed packaging puts food at risk from microbial contamination – the effectiveness of seals must be checked.
- If using glass jars or bottles, a glass-breakage action plan should be in place – glass can cause serious injury if eaten.
- Low-acid (above pH 4.6) vacuum or modified-atmosphere packaged foods must have specific control factors applied to reduce the risk of growth and toxin production by *Clostridium botulinum*.
- Use-by dates must be used on products if there is:
  - a possibility pathogenic microorganisms could be present, and
  - these pathogens could grow or produce toxin to unacceptable levels in the product, at the intended storage temperature.

  These are requirements of the Code.
- Mandatory statements and declarations must be present on product labels if required by the Code; for example, the presence of food allergens must be declared.
- VITAL is a software tool that provides assistance when determining if the presence of an allergen needs to be declared on products.
- There are specific labelling requirements in the Code for several types of food including baby food, baby formula and fermented smallgoods; these must be followed.
- Information allowing products to be traced back to the business that made them, and to a specific batch, must be provided on labels. This is a requirement of the Code.
- Information on labels must be easy to read, particularly if related to food safety; for example, the size of printing required for warning statements is specified in the Code.
- Directions for use and/or directions for storage of food must be included on the label if required for reasons of health and safety. This is a requirement of the Code.

FOOD STANDARDS
Australia New Zealand

Food Industry Recall Protocol

A GUIDE TO CONDUCTING A FOOD RECALL AND WRITING A FOOD RECALL PLAN

6th Edition September 2008

FOOD RECALL

Farmer Nicks Organics
Wholemeal Flour – Gluten Free
500 g plastic bag
Best-before: September 2010

Farmer Nicks Organics Pty Ltd is conducting a voluntary consumer level recall on the ...ve product due to an incorrect gluten free ... above batch was tested and found ...h levels of gluten.

... a gluten allergy ... consume this ...erned about their ...edical advice.

...to the above product ...e and best-before date. ...s Organics products are ...by this recall.

...elling irregularity there is ...n this product. Consumers ...tolerant to gluten can safely ...ume this product.

...who have a gluten allergy or ...e asked to return the product to ... of purchase for a full refund.

...ologise for any inconvenience.

...mer Nicks Organics Pty Ltd
...Wentworh Rd Griffith NSW 2680
(02) 6969 1234

FOOD RECALL

Kathy's Kitchen
Kids Health Bars
40 g plastic packet
Best-before date: 1 Jul 2010

Kathy's Kitchen Pty Ltd is conducting a voluntary consumer level recall of the above product due to the possibility that it may contain small glass fragments.

Customers should not consume the product. Any consumers concerned about their health should seek medical advice.

The recall applies only to the above product with the nominated size and best-before date. No other Kathy's Kitchen products are affected by this recall. Customers are asked to return the product to their place of purchase for immediate full cash refund.

We apologise for any inconvenience.

Kathy's Kitchen Pty Ltd

FOOD RECALL

Swan's Seafoods
Salmon Dip
250 g plastic tub
Use-by date: 3 Sep 2009

Swan's Seafoods Pty Ltd is conducting a voluntary consumer level recall of the above product in response to testing which indicates the presence of Listeria monocytogenes bacteria.

Listeria may cause illness in pregnant women, the very young, the elderly and people with low immune systems. Any consumers concerned about their health should consult their doctor.

Customers should not consume the product. They should return it to the place of purchase for a full refund.

The recall applies only to the above product with the nominated size and use-by date. No other Swan's Seafoods products are affected by this recall.

We apologise for any inconvenience.

For further information please call:

(03) 6233 1234

Swan's Seafoods Pty Ltd

3 Penny Lne Hobart TAS 7000

FOOD STANDARDS
Australia New Zealand
Te Mana Kounga Kai – Ahitereiria me Aotearoa

Chapter 8

What if something goes wrong?

Food products that are determined to be a potential food safety risk to consumers must be removed from the marketplace as quickly as possible. This is achieved by performing a food recall.

This chapter guides you through the process of writing a food recall plan for your business. The information provided here is from the Food Industry Recall Protocol: A Guide to Writing a Food Recall Plan and Conducting a Food Recall produced by FSANZ (see page 271). Before writing or reviewing your plan, it is recommended you obtain this FSANZ guide because it contains additional information that you will require.

Food recalls – overview and definitions

Before providing specific guidance on writing a food recall plan, here is some key information about the process and requirements.

Reasons for food recalls

A recall is required to remove from distribution, sale or the risk of being eaten, food that may pose a health and safety risk to consumers.

In general, food becomes unsafe because something has 'gone wrong'; examples include:

- An allergen has been added to a product but its presence has not been declared on the label.
- Potentially hazardous foods have been stored above 5°C for sufficient time to allow any pathogens that may be present to grow and/or produce toxins to unacceptable levels.
- A temperature probe failed and a product batch was inadequately heat processed (under-processed).
- A harmful chemical was mistaken for an ingredient and was added to a product batch.

- A potentially harmful level of a permitted food additive, such as a preservative, was mistakenly used.
- The pH-adjusting step of the production process was accidentally missed out.
- Shards of glass from an undetected breakage on the processing line have contaminated products.
- A packaging supplier has notified its customers that a batch of plastic pouches used for heat processing in-pack has faulty seals.
- An incorrect use-by date was stamped on product labels.

Legal obligations

It is a requirement of the Code for manufacturers, wholesalers and importers of food to have a documented food recall plan in place to ensure the effectiveness of recalls of potentially unsafe food (Standard 3.2.2). The information provided here is only a guide and you should seek independent advice on whether your recall plan will comply with all requirements from a technical expert or legal adviser.

Recall plans must be supplied to an authorised officer from a federal, state or territory authority, on request. If a recall needs to be initiated, the business must follow the procedure specified in its recall plan.

Food recall plans

Food recall plans are documents that provide details of the system a food business will use if it needs to retrieve products or notify those who may have its products that they must be disposed of.

The plan must be written specifically for each individual business because it needs to be relevant for the business's product range and distribution networks; a 'generic' document cannot simply be copied.

Recall plans should be detailed and structured in a way that key information can be found quickly. Staff who will be involved in implementing a recall must have a sound understanding of their responsibilities as outlined in the plan.

Initiation of recalls

Food businesses may need to seek expert assistance to determine if a recall is necessary. Therefore it is important to quickly gather as much information as possible about the nature of the potential problem so the assessment can be made.

Recalls can be initiated in two ways: voluntarily or mandated by government authorities.

Voluntary recalls

A voluntary recall is initiated by the food business that has primary responsibility for the supply of the affected product. This action must be taken if the business becomes aware that there is a potential food safety issue with a product – either discovered by the business or a third party.

Examples may include:

- Food business – product testing results from a laboratory have come back positive for the presence of pathogenic bacteria and the product has failed to meet the microbiological limits criteria specified in the Code (Standard 1.6.1).
- Manufacturer – an ingredients supplier has identified that a batch of ingredients it has supplied to the business is unsafe.
- Wholesalers – despite storage at the correct temperature, swelling of packs has been observed before the use-by date has passed.
- Retailers – complaints from consumers to whom they have sold the product have been received, and it is confirmed that the complaint is genuine (e.g. the presence of glass fragments).
- Consumers – they call or write directly to the business to complain, and it is confirmed that the complaint is genuine (e.g. the presence of glass fragments).
- Government agencies – which act in response to reports they have received and then verified as correct.

The use of the word 'voluntary' does not mean that those involved in the products storage, distribution and sale can chose whether or not to return or dispose of the product. All of the products affected by the recall must be removed from the market place immediately.

Mandatory recalls

Mandatory recalls can be initiated by Australian federal, state or territory authorities. This action is taken if there is a serious public health and safety risk. For example, a foodborne illness outbreak occurs, but the manufacturer of the implicated product fails to take adequate steps to protect consumer safety.

Levels of recalls

Recalls can either be performed at the trade level, or at both the trade and consumer level.

Trade recalls

A trade recall is required when the affected products have not been made available for direct purchase by the general public.

Trade recalls involve:

- distribution centres
- wholesalers
- food service outlets
- caterers.

Consumer recalls

A consumer recall is more extensive than a trade recall. It involves recovery of or disposal of products from every stage of the storage and distribution chain, including consumers.

Consumer recalls involve:

- trade outlets (as listed above)
- retail outlets, such as
  - supermarkets
  - convenience stores
  - delicatessens
  - health food stores
  - fruit and vegetable shops
- homes of consumers.

Priorities in the recall process

These tasks must be the focus at the start of a recall:

- Stop the distribution and sale of the affected product immediately.
- Notify government authorities (trade and consumer recalls) and the public (consumer recalls only).
- Effectively and efficiently remove from sale any product that is potentially unsafe.

Other important priorities are:

- gathering relevant information about the product
- determining if a trade or consumer-level recall is needed
- notifying other relevant parties
- retrieving the affected products
- disposing of the affected products

- determining what actions may be required to prevent re-occurrence of the problem
- reporting the outcomes of the recall to the relevant authorities.

Notification and reporting requirements

One of the key responsibilities of a food business performing a recall is to notify and report details of the recall to various authorities and bodies.

Government authorities

Home State or Territory Action Officers – the state or territory where the head office of the business that manufactured or imported the affected food product is called the Home State or Territory. Home State or Territory Action Officers are responsible for providing advice on whether a recall is required and at what level, in addition to liaison with FSANZ. They will also work with the food business to determine how the affected products will be collected and disposed of, and what corrective actions must be implemented to prevent re-occurrence. The Action Officer must be contacted by telephone as soon as a food business suspects a recall may be required; numbers are provided in the FSANZ Protocol.

FSANZ – FSANZ coordinates the recall action and liaises with the food business, commonwealth and state and territory authorities. FSANZ recall coordinators should be contacted by telephone upon initiation of a recall; numbers are provided in the FSANZ Protocol.

Consumer affairs – Commonwealth Ministers responsible for consumer affairs must be notified in writing within 2 days of initiating the recall. State or territory ministers responsible for consumer affairs may also need to be notified. The FSANZ recall coordinator can perform these notifications on behalf of a food business. However, details on how to notify relevant ministers are provided in the FSANZ Protocol.

Distribution networks (trade and retail level)

Distribution networks must be notified so that the distribution of the affected products can be prevented or halted. This must be done by telephone as soon as a decision has been made. This should be followed up in writing, and the FSANZ recall coordinator will assist you to do this. Additionally, template letters are provided in the FSANZ Protocol. Notification should not be provided in written form alone because mail and email addresses can change and the information may not be received.

The public

The public must be notified if a consumer-level recall is initiated; that is, if the products have been made available for sale on the retail market.

Methods to inform the public vary according to the size and complexity of the distribution network; examples include:

- newspaper advertisements (including online)
- media releases
- displaying signs in retail outlets where the products are sold.

A newspaper advertisement is the most common method used to notify consumers.

The FSANZ recall coordinator and the Home State or Territory Action Officer can help you chose the appropriate communication method(s) required. Specific requirements to ensure recall advertisements are clearly visible must be followed; see Box 78 and the FSANZ Protocol for further information.

Food industry organisations

Organisations that can assist you to notify your distribution network, or that need to be made aware of the issue, should also be notified. For example, if products are distributed to school tuck shops, the Independent School Canteen Association should be notified.

Box 78 – Example food recall advertisement

Example Only FOOD RECALL (1)

Pete's Pasta Products
Fresh filled Cannelloni Pasta – 500g
Use by date: 25/06/09 (2)

Pete's Pasta Products Pty Ltd is conducting a voluntary consumer level recall of the above product due to microbial contamination. (3)

Tests have detected abnormally high levels of E.coli bacteria. Customers should not consume the product. Any consumers concerned about their health should seek medical advice. (4)

The recall applies only to the above product with the nominated size and use by date. No other Pete's Pasta Products are affected by this recall. Customers are asked to return the product to the point of purchase for immediate full cash refund. (5 & 6)

Pete's Pasta Products is greatly concerned at any risk to its customers. This recall is being undertaken to ensure the safety of our customers as an ongoing commitment to maintain the highest possible standards of safety and product quality at all times.

We apologise for any inconvenience.

For further information please call:

1800 808 966

Pete's Pasta Products Pty Ltd.

13 Wattle Avenue Canberra ACT 2600 (7)

1 – Type of recall

2 – Name, size and description of product

3 – Reason for recall

4 – Hazard

5 – Identify

6 – Disposal

7 – Company contact details

Source: Food Industry Recall Protocol: A Guide to Writing A Food Recall Plan and Conducting a Food Recall

Post-recall reporting

After the recall is completed, a post-recall report must be provided to FSANZ. Details required include the amount of product recalled compared with the amount produced, and the corrective actions that will be implemented to prevent further re-occurrence of the problem that caused the recall. Details on the other information that needs to be included is provided in the FSANZ Protocol.

An effective recall requires fast facts

Information that clearly identifies, and therefore allows the tracing of, the affected product must be provided quickly so the recall can be effective. The food business must supply specific details about the affected product. These are the:

- nature of the problem
- brand name and description, including package size and type
- use-by or best-before date
- lot identification (batch or serial number)
- quantity of the batch made, the date and quantity released
- quantity that can be accounted for
- distribution within Australia, including a distribution list and the types of premises at which the product is likely to be sold
- overseas distribution (if applicable)
- importer information (if imported).

A photo or other image to help identify the product may also be of use.

Retrieval of recalled products

Products are either returned via their distribution network, by retailers or directly by consumers. It is usual practice for consumers to return products to their place of purchase, where they are refunded the purchase price.

Some major retailers will retrieve and dispose of recalled products at store level.

To track how much food is recovered for comparison against how much was distributed or sold, accurate records must be collected: tallying the amount of product recovered.

The fate of recalled products

It is a requirement of the Code that a food business makes sure food that is recalled is held, separated and is clearly distinguished from non-affected food until it is:

- destroyed or disposed of – for example, removed from its packaging and placed in landfill. An Environmental Health Officer may be required to oversee this procedure
- used for purposes other than food for humans – for example, as animal feed or fertiliser
- returned to its supplier – for example, if the product is at a retailer who is holding it until the supplier comes to pick it up
- further processed in a way that ensures its safety and suitability – for example, heat processed using a time–temperature combination adequate to kill a contaminating pathogen. Permission from the Home State or Territory Action Officer must be received and expert advice must be sought before using this option
- determined to be safe and suitable – for example, the products may have been recalled because it was suspected they may have been contaminated with pathogenic bacteria, but testing by a laboratory determines that this is not the case.

Decisions about the method of disposal or modification of recovered recalled products must be made in conjunction with the Home State or Territory Action Officer.

Writing a recall plan

A recall plan must contain the complete details of a businesses food recall system. This includes:

- staff responsibilities and procedures they need to follow
- a list of who to notify, including their contact details
- distribution network information
- plans for how products will be retrieved
- information on how the success of the recall will be assessed.

Roles and responsibilities

All tasks that need to be performed for a food recall must be assigned to the business owners or specific staff members. The responsibilities of those assigned roles in a recall must be clearly defined. In small businesses, individuals may have to perform multiple roles.

Some of the responsibilities of those involved in managing a recall include:

- liaison with government authorities, including notification and reporting
- coordinating retrieval and disposal of products
- preparation of press advertisements to notify consumers (if needed)
- providing details of the distribution networks of affected products, such as a distribution register (see Box 79)

- determining and implementing corrective actions
- preparing post-recall reports.

Staff members given responsibilities in a food recall should be provided with training on the recall system.

Some tasks may be performed by external consultants, such as employing an agency to prepare press releases to notify consumers.

Notification procedures and contact details

Government authorities

A list of authorities to notify should be included in the recall plan. Contact details for these authorities are provided in the FSANZ Protocol.

Distributors, wholesalers, retailers

All businesses that are part of the distribution network, or that are selling the affected products, must be notified of the recall and product details. Full contact details should be included in the plan and these should be kept up to date.

Information that needs to be provided is:

- the name of the product and the batch code, use-by or best-before date
- the reason for the recall
- where to return unsold products
- who to contact for further information.

It is important to instruct your distributors to tell their customers as soon as you notify them of the recall. The faster the word is spread throughout the entire distribution chain, the less likely that someone will become ill from eating an affected product.

The public

If a consumer-level recall is initiated then the public must be notified.

The recall plan should provide details of the process that will be used to notify the public, including the person responsible for making the arrangements. Unless all products are distributed and sold using the same network, separate information will need to be included in the plan for each different product type.

Box 79 – Example recall distribution register

RECALL DISTRIBUTION REGISTER					
Food product name			Use-by date/batch code etc.		
Product size			Date of manufacture		
Total amount produced (kg/cartons)					
Distribution profile (from despatch records)					
Date	Customer name	State	Amount sent to customer	Date customer contacted about recall	Amount of stock remaining
Total stock accounted for =					
Total stock unaccounted for (e.g. Stock in trade/consumed)					

Source: Food Industry Recall Protocol: A Guide to Writing a Food Recall Plan and Conducting a Food Recall

Contact details for relevant newspapers and other media networks, and draft examples of recall notices, can be included in this section of the plan.

Access to production and distribution records

The recall plan should provide information on how to determine the quantity of specific batches of product made and where they have been distributed. It should also state who is responsible for compiling this information if a recall is initiated.

Instructions for how to quickly locate important records should be included in the plan:

- a complete and up-to-date history for all product batches, including which batches of ingredients and packaging materials were used for each
- the total quantity of specific product batches manufactured
- those that track the use and disposal of all raw ingredients and packaging materials

- adequate details of the customers in the distribution network to which specific product batches have been sold or distributed.

Rather than printing off standard stock control or financial reports from your database, a specific recall report should be set up, which enables you to quickly sort distribution data.

Product retrieval and disposal

The recall plan should include details on the method for retrieving returned products. This should incorporate the process for return via retailers, distribution chains or directly from consumers. The name or names of staff responsible for coordinating the process should be specified in the plan.

It is important that the amount of the product returned can be compared with the amount distributed. Including a recall distribution register, such as the one shown in Box 79, in the recall plan can assist with this procedure.

The plan should state the system that will be used to separate and store returned products. This would detail labelling methods to clearly identify the affected items and where they should be held.

Details of product disposal do not need to be specified in the plan as this is determined on a case-by-case basis in consultation with the Home State or Territory Action Officer.

Trialling and reviewing the recall plan

To be certain that everyone knows what their role is in the event of a recall, and that your record keeping systems are sufficient, it is recommended that you perform mock trials on your recall plan. This should simulate the recall of current food products, which can be likened to running a trial fire evacuation to prepare for a real emergency. This can be used to highlight any weaknesses in your recall plan that need to be rectified. It is critical that all involved parties are advised that you are performing a mock recall and not an actual recall.

The recall plan should also be reviewed regularly, particularly if:

- changes are made to the product range manufactured
- there are staffing or responsibility changes
- changes are made to your distribution networks
- changes are made to your record keeping systems
- there are changes to the contact details for customers, or others who require notification
- FSANZ or state or territory authorities modify their requirements
- a trial recall is performed.

KEY MESSAGES FROM CHAPTER 8

- A recall removes food that may pose a health and safety risk to consumers from the marketplace.
- Manufacturers, wholesalers and importers of food must have a documented food recall plan; this is a requirement of the Code.
- Recalls can either be initiated voluntarily or they can be mandatory.
- Recalls can be performed at the trade or consumer level.
- The main priorities at the start of a recall are:
  - stopping the distribution and sale of the affected product
  - notifying the relevant parties including government authorities and the public
  - removing from sale any product that is potentially unsafe.
- FSANZ recall coordinators and Home State or Territory Action Officers assist food businesses with food recalls.
- The ability to identify and trace the distribution of affected products are key requirements of an effective recall.
- Food that is returned to a business during a recall must be held, separated and clearly distinguished from non-affected food until it is deemed to be safe, destroyed or disposed of. This is a requirement of the Code.
- The recall plan should include details of staff responsible for performing a recall and their roles.
- Recall plans should be tested (mock trials) and reviewed regularly.

Chapter 9

Food microbiology: further facts

Overview of pathogenic bacteria that cause infections

The table below provides a summary of the major bacterial pathogens that cause foodborne infections and illnesses.

The symptoms listed are for primary illnesses; other secondary illnesses, which are usually longer term, may develop following the initial illness. Unless otherwise noted, the primary illnesses are different forms of gastroenteritis. Secondary illnesses for some individual bacteria will be discussed later in this chapter.

Illnesses (infection) caused by pathogenic bacteria

Bacteria	Typical incubation period	Symptoms
Aeromonas species	1 to 2 days	Watery diarrhoea, abdominal pain, nausea
Campylobacter species	3 to 5 days	Abdominal pain, diarrhoea (often containing blood), fever
Escherichia coli O157 and related types (Box 80)	2 to 5 days	Watery diarrhoea, followed by bloody diarrhoea (haemorrhagic colitis), severe abdominal pain, bloody urine
Escherichia coli other pathogenic types	0.5 to 3 days	Profuse watery diarrhoea, abdominal pain, vomiting, dehydration
Listeria monocytogenes	4 to 21 days	Listeriosis: flu-like symptoms, meningitis, septicaemia, encephalitis, spontaneous abortion, stillbirths

Bacteria	Typical incubation period	Symptoms
Salmonella species (Box 81)	18 to 36 hours	Abdominal pain, diarrhoea, fever, vomiting, septicaemia
Shigella species	1 to 3 days	Watery diarrhoea (may contain blood, pus and mucus), abdominal pain, vomiting
Vibrio cholerae	2 to 3 days	Profuse watery diarrhoea, vomiting, dehydration, abnormal acid levels in blood
Vibrio parahaemolyticus	12 hours	Abdominal pain, diarrhoea, vomiting, fever
Vibrio vulnificus	38 hours	Skin infections, gastroenteritis (fever, chills, nausea), septicaemia
Yersinia enterocolitica	1 to 7 days	Abdominal pain, fever, watery diarrhoea

Box 80 – What does the 'O157' mean when written after *Escherichia coli*?

All humans, and many animals, naturally carry large numbers of harmless non-pathogenic *E. coli* in their normal intestinal flora. However, there are some other types of *E. coli* that can cause mild, or even severe, illness if eaten. The potentially pathogenic types can be separated from non-pathogenic types by testing for specific traits (e.g. which toxins they can produce).

When *E. coli* is implicated in an outbreak of foodborne illness, it is important to identify which type has caused the outbreak, so authorities can then match the type isolated from those affected with any found in leftover portions of food these people have eaten. This process assists authorities to identify the source of contamination, much like a detective uses DNA identification in a murder investigation. It is also very useful for doctors treating those affected to know with which type of *E. coli* the patients are infected. The 'O157', for example, indicates a particular type of *E. coli* that can cause severe long-term health problems.

Overview of pathogenic bacteria that cause intoxications

The table below provides a summary of the different bacterial pathogens that cause foodborne intoxications.

Illnesses (intoxication) caused by pathogenic bacteria

Bacteria that form toxins	Toxin type	Typical incubation period	Illness and symptoms
Bacillus cereus	Emetic	0.5 to 5 hours	Gastroenteritis: nausea, vomiting
	Diarrhoeal	8 to 16 hours	Gastroenteritis: profuse watery diarrhoea, abdominal cramping
Clostridium botulinum	Neurotoxin	18 to 36 hours	Botulism: dizziness; eyesight problems; difficulty swallowing, speaking and breathing; muscular weakness; respiratory paralysis; death
Clostridium perfringens	Enterotoxin	8 to 22 hours	Gastroenteritis: abdominal pain and diarrhoea
Staphylococcus aureus	Enterotoxin	2 to 4 hours	Intoxication: nausea, vomiting, abdominal pain, diarrhoea

Box 81 – *Salmonella* Typhi

One type of *Salmonella*, *Salmonella* Typhi, causes a severe illness called typhoid fever. This is predominantly a problem in developing countries and most Australian cases are in individuals who have returned from overseas travel (particularly Asia). Foods typically involved are fresh fruit and vegetables, milk and shellfish. Contamination of these foods is either through unhygienic handling by people who have typhoid fever, or by use of untreated water when preparing foods (e.g. washing salad items).

Australian food handlers who suspect they have been infected with *Salmonella* Typhi, or have had close contact with a person who has been diagnosed with typhoid fever, must seek medical advice before continuing their normal work duties.

Individual pathogenic bacteria – up close and personal

Aeromonas species

- Illness caused:
  - Gastroenteritis that usually lasts for 1 to 7 days.
- Growth conditions:
  - Approximately 30°C optimum, although cells can grow slowly at refrigeration temperatures.
- Specific control measures:
  - Cook seafood using adequate time–temperature combinations, such as 70°C for 2 minutes or equivalent.
  - Prevent juices from raw seafood contaminating ready-to-eat foods.

Bacillus cereus (spore-forming)

- Illness caused:
  - Gastroenteritis, with the specific symptoms dependent on which toxin type is produced.
- Growth conditions:
  - Some types can grow at 5°C, but only slowly.
  - The optimum growth temperature is 30–37°C.
- Spores can survive heat processes that kill any cells present in the food. *B. cereus* prefers to grow in the absence of other competing microorganisms, which means it can increase to high levels unless food is chilled appropriately after cooking or is eaten soon after cooking. See Box 82 for information about how the failure to follow either of these steps during the preparation of fried rice can cause foodborne illness.
- Specific control measures:
  - Rapidly cool food after cooking to 5°C or below, or maintain food at 60°C or above until eaten.
  - Adequately acidify rice used to prepare sushi; the NSW Food Authority has published specific guidance for this: Food safety guidelines for the preparation and display of sushi (see page 272).
- The preservative nisin can also be used as a control measure. It acts by keeping spores in a dormant state, preventing cells forming and producing toxin. Nisin can only be used in certain food types, as specified in the Code.

Box 82 – *Bacillus cereus* and fried rice

Foodborne illness caused by *B. cereus* was often associated with consumption of fried rice.

The first stage in preparing fried rice is to cook the rice by boiling or steaming. Low levels of *B. cereus* spores are commonly found in raw rice and this heat process is not severe enough to kill all of the spores that may be present. Rather, the heat process can activate dormant spores so they form cells that are then able to grow and produce toxin.

When chilled, rice tends to clump, so it was common practice to leave it at room temperature until required. However, leaving the cooked rice at room temperature for too long, or cooling it too slowly between the boiling and the frying step, can give the cells enough time to produce toxin to high enough levels to cause illness. *B. cereus* toxin is sufficiently heat resistant to survive the frying process.

Campylobacter species

- Illnesses caused:
  - Most common cause of acute gastroenteritis, called campylobacteriosis, in Australia.
  - In 2007, nearly 17 000 campylobacteriosis cases were reported, but these figures do not include all cases from NSW where only outbreaks (not isolated cases) of campylobacteriosis require reporting.
  - Campylobacteriosis can cause severe bloody diarrhoea, but symptoms can also be so mild they may go unnoticed or be misdiagnosed; these cases will then also be left out of reported figures.
  - Development of a serious, longer term illness called Guillain-Barré syndrome can occur in a small proportion of people following infection with *Campylobacter.*
  - Guillain-Barré syndrome causes gradual degradation of the nervous system, which can lead to loss of mobility and may cause breathing difficulties.
- Growth conditions:
  - Optimum growth temperature is 42°C, which is higher than that of most other non-spore-forming pathogenic bacteria.
  - *Campylobacter* does not grow well in foods, but it can survive.

- *Campylobacter* does not have to grow in food to be hazardous; as few as 500 cells per gram of food may cause illness.
- Specific control measures:
  - Cook all raw meats using adequate time–temperature combinations, such as 70°C for 2 minutes or equivalent.
  - Prevent cross contamination from raw chicken or red meat to ready-to-eat foods such as salads.
  - Prevent food handlers from performing normal duties if suffering from gastroenteritis.
  - Make sure that food handlers maintain appropriate levels of personal hygiene (e.g. washing hands thoroughly after going to the toilet).

Clostridium botulinum (spore-forming)

- Illness caused:
  - Botulism intoxication affects the nervous system, particularly nerves that control muscles. If left untreated, botulism can be fatal as it eventually paralyses the muscles needed for breathing. Even if early treatment occurs, patients can still spend months in hospital.
  - The above description is for the classical type of botulism where the toxin, formed during cell growth in food, is eaten. There is another type called infant botulism (Box 83).
- Growth conditions:
  - Some types of *C. botulinum* cells can grow at refrigeration temperatures, down to 3°C.
  - Cells prefer to grow and produce toxin in the absence of oxygen. This means that some packaging or processing techniques may be more hazardous. For example, vacuum packaging removes the air, including oxygen, from around foods. The oil in products such as preserved garlic protects any *C. botulinum* present from oxygen in the air. Even foods that do not fit into these two categories can still support *C. botulinum* growth because oxygen-deficient pockets can be created within food when it is processed or packaged.
- *C. botulinum* has the most heat resistant spores, compared with *B. cereus* and *C. perfringens.*
- Botulism is a very rare illness in Australia. In the northern hemisphere, cases are reported more often because home preserving (e.g. canning, bottling, salami manufacture and smoking) of foods such as vegetables, meats and fish is more common.

Box 83 – Botulism, babies and honey

The Code specifies that commercially manufactured infant food cannot contain honey unless the honey has been processed to kill *Clostridium botulinum* spores (Standard 2.9.2). This special precaution is in place because there is a form of botulism, known as infant botulism, which has been linked to honey. This illness is caused when dormant *C. botulinum* spores are eaten, become active and form cells in the gastrointestinal system, grow and produce toxin. Babies less than 1 year old are most likely to suffer from infant botulism; adults are rarely affected. Symptoms typically start with constipation followed by lethargy, listlessness, poor feeding, drooping eyelids, difficulty swallowing, and overall muscle weakness (often described as 'floppy baby').

- Specific control measures:
  - Heat food in hermetically sealed containers using validated time–temperature combinations designed to kill virtually all target cells and spores (i.e. conventional canning process).
  - Heat food using less harsh time–temperature combinations (e.g. 90°C for 10 minutes), in conjunction with recipe control and/or product storage at 5°C or below for a limited time. Reducing the pH to below 4.6 is one example of recipe control.
- The preservatives sodium nitrite and nisin can be used to control growth in certain food types, as specified in the Code.

Clostridium perfringens (spore-forming)

- Illness caused:
  - Gastroenteritis caused by the toxin is generally not life threatening and lasts for only about 24 hours.
- Growth conditions:
  - Cells cannot grow in the presence of oxygen or at temperatures below about 15°C.
  - The optimum temperature for growth is around 44°C.
- *C. perfringens* levels present in raw ingredients will usually be much lower than the level required to cause illness, which is at least one million cells per gram of food. Therefore, cells have to grow in food to cause illness.

- Outbreaks involving large groups of people are relatively common. They are often linked to food service preparation of meals for large numbers of people, such as for catered events or at restaurants.
- Specific control measures:
  - Rapidly cool food after cooking to 5°C or below, or maintain foods at 60°C or above until eaten.
  - Re-heat food to above 75°C throughout before serving.

Escherichia coli O157 and related types

- Illnesses caused:
  - Gastroenteritis of varying duration and severity.
  - Some affected people develop more severe symptoms with serious long-term complications that may be fatal (particularly in vulnerable populations); for example, *E. coli* O157 may cause haemorrhagic colitis, characterised by the presence of bloody diarrhoea.
  - Up to 30% of people who develop haemorrhagic colitis then develop haemolytic uraemic syndrome (Box 84).
- Does not have to grow in food to be hazardous; as few as 10 cells are required to cause illness.

Box 84 – Haemolytic uraemic syndrome (HUS)

HUS can develop about a week after the onset of diarrhoea caused by *E. coli* O157. Children are particularly susceptible to HUS illness.

A toxin enters the bloodstream after which it can damage the kidneys and blood cells. Kidney failure can occur and then dialysis treatment is required. Death occurs in 8–10% of cases, commonly those who are in vulnerable populations.

Chronic high blood pressure or end-stage kidney failure, requiring long-term dialysis or kidney transplantation, may develop in patients who recover from the initial acute phase of HUS. HUS can also cause diabetes, resulting in life-long dependence on insulin treatment.

- Specific control measures:
  - Use adequate time–temperature combinations (e.g. 70°C for 2 minutes) to cook meat, particularly boned and rolled portions, minced meat or products made from minced meat (e.g. sausages).
  - Prevent raw meat juices contaminating any ready-to-eat foods.
  - Follow recommended recipe and processing guidelines for uncooked fermented meats (smallgoods).
  - Do not use salad vegetables that have been fertilised with animal manure or fallen fruit from orchards where animals have been allowed to graze.

Listeria monocytogenes

- Illnesses caused:
  - Listeriosis, which can result in mild, flu-like symptoms in non-vulnerable populations.
  - Pregnant women are at particular risk from listeriosis, which can cause spontaneous abortions or stillbirths.
  - *L. monocytogenes* can also cause severe illness such as meningitis and septicaemia
  - Vulnerable populations, including pregnant women and the elderly, have a much higher chance of becoming severely ill and dying if infected.
  - All those at higher risk should avoid eating foods that are likely to contain *L. monocytogenes* (Box 85).
- Growth conditions:
  - Can grow at refrigeration temperatures, even below 1°C.
  - Can grow in foods that are vacuum packed because it does not require oxygen to grow.
- Specific control measures:
  - Cook foods using adequate time–temperature combinations (e.g. 70°C for 2 minutes) and avoid re-contamination after heating.
  - If preparing foods that have no heating step, strictly follow hygienic practices and make sure a very robust cleaning and sanitation program is used.
- Guidance for *L. monocytogenes* control in specific food industry sectors have been published, including the NSW Food Authority's *Listeria* Management Program (for businesses manufacturing packaged ready-to-eat meat products) and The Australian Manual for Control of *Listeria* in the Dairy Industry (Australian Dairy Authorities' Standards Committee). Details about these are provided at the end of the book.

Box 85 – *Listeria monocytogenes*: foods to avoid for those at risk

The following table lists the foods that are more likely to be contaminated with *L. monocytogenes*; these foods should not be eaten by people who are vulnerable to foodborne illness.

Examples of foods at higher risk of causing listeriosis

Food type	Examples
Cold meats	Unpackaged ready-to-eat from delicatessen counters, sandwich bars, etc. Packaged, sliced ready-to-eat
Cold cooked chicken	Purchased (whole, portions, or diced) ready-to-eat
Pâté	Refrigerated pâté or meat spreads
Salads (fruit and vegetables)	Pre-prepared or pre-packaged salads from salad bars, smorgasbords, etc.
Cold seafood	Raw (e.g. oysters, sashimi or sushi) Smoked ready-to-eat Ready-to-eat peeled prawns (cooked) e.g. in prawn cocktails
Cheese	Soft, semi-soft and surface ripened cheeses (pre-packaged and delicatessen) e.g. brie, camembert, ricotta, feta and blue
Ice cream	Soft serve
Other dairy products	Unpasteurised dairy products (e.g. raw goats milk)

Source: adapted from *Listeria* and food – advice for people at risk (see page 273).

Salmonella species (non Typhi)

- Illnesses caused:
  - Infection causes mild to severe gastroenteritis, called salmonellosis.
  - Septicaemia may occur in some patients.
  - Reactive arthritis (also known as Reiter's syndrome) – an illness that causes debilitating pain in large joints (lasting for weeks or months) – can develop in a small percentage of people after recovery from salmonellosis.

- Dishes containing raw or undercooked eggs prepared in private homes or in food service settings have been implicated in foodborne illness outbreaks. Individuals in vulnerable populations should avoid eating fresh eggs unless they are thoroughly cooked.
- Salmonellosis outbreaks have been linked to foods that most people would not have considered to be hazardous (Box 86).
- Growth conditions:
  - *Salmonella* cannot usually grow below 7°C.
  - The optimum growth temperature is 35–43°C.
- In some foods, especially high-fat, low-moisture foods, fewer than 100 cells may cause illness; therefore growth of *Salmonella* is not always necessary to cause illness.
- Specific control measures:
  - Cook meat, particularly poultry (whole chickens or pieces), using adequate time–temperature combinations (e.g. 70°C for 2 minutes), and avoid re-contamination after heating.
  - Use commercially processed pasteurised egg products (available in whites, yolk and whole egg) where possible. If using fresh eggs, avoid using those with cracked or dirty shells because these are more likely to be contaminated. Store fresh eggs at 5°C or below.

Box 86 – Peanut butter, chocolate and cheese implicated in salmonellosis outbreaks

Contamination of commercially manufactured peanut butter, chocolate and cheese products with *Salmonella* has been linked to salmonellosis outbreaks affecting large numbers of people:

- peanut butter (United States, two outbreaks 2006 and 2008/9) – over 400 people each outbreak
- chocolate (United Kingdom, 2006) – 37 people
- chocolate (Europe, 2001–2002) – 439 people
- peanut butter (Australia, 1996) – 51 people
- cheese (Canada, 1994) – over 1500 people.

These foods are all relatively high in fat and low in available moisture. These characteristics protect *Salmonella* cells from any heat processes used during manufacturing (e.g. chocolate conching). Additionally, the fat in the products protects bacterial cells against acidic conditions in the human digestive system. This means that very low levels of contamination are capable of causing illness.

Food businesses that prepare these products, or those with similar characteristics, should not be too relaxed about food safety because they believe they are low risk. Primary raw ingredients used in these products should have received a heat process able to kill all pathogenic bacteria that may cause a problem. These processes include roasting peanuts, roasting cocoa beans and pasteurising milk.

Controls must be in place to prevent recontamination of these ingredients with *Salmonella* or other pathogens. Incorrect layout of premises, poor pest-control practices and cross contamination of heat processed ingredients by non-heated ingredients are the key risk factors.

Shigella species

- Illness caused:
  - Infection is called shigellosis and is a form of gastroenteritis associated with diarrhoea containing blood.
  - Rarely, other illnesses may develop after the initial bout of gastroenteritis; these include HUS, reactive arthritis and septicaemia.
- Growth conditions:
  - *Shigella* cannot grow below 10°C or in acidic foods.
- Only low levels of cells, 10 to 100 in total, are required to cause illness.
- Specific control measures:
  - Cook foods using adequate time–temperature combinations (e.g. 70°C for 2 minutes) and avoid re-contamination after heating.
  - Prevent food handlers from performing normal duties if suffering from gastroenteritis.
  - Make sure food handlers maintain appropriate levels of personal hygiene (e.g. washing hands thoroughly after going to the toilet).

Staphylococcus aureus

- Illness caused:
  - Some types of *S. aureus* produce toxins as they grow in food; these can cause illness (gastroenteritis).

- Symptoms can start as soon as 1 hour after eating the toxin and generally last 24 hours or less.
- The short duration of the illness means that those affected frequently do not visit a doctor, and the number of cases reported is likely to be artificially low.

- Growth conditions:
  - Cells can grow in foods that have lower water activity and higher levels of salt or sugar, compared with other foodborne pathogenic bacteria.
  - Although toxin can be produced in food kept between 10–46°C, it is produced more readily above 40°C.
- *S. aureus* needs to grow to relatively high levels for enough toxin to be produced to cause illness. The toxin will remain active in food even after it is cooked.
- *S. aureus* is naturally present in the nose of about 50% of the population and is also often present on the skin. *S. aureus* is a common cause of infection of skin wounds.
- *S. aureus* foodborne illness outbreaks commonly involve food that is handled after cooking and then left unrefrigerated for over 4 hours. One example is preparation of chicken sandwiches. If the chicken is cooked properly, all pathogenic bacterial cells will be killed. However, if it is then touched by hands contaminated with *S. aureus* and left at warm temperatures, the few bacteria transferred to the chicken will be able to grow readily because there are no other microorganisms present to compete with them.
- Specific control measures:
  - Make sure food handlers do not handle ready-to-eat food with unwashed hands, particularly if they have any infected skin wounds.
  - Rapidly chill food to 5°C or below, or keep food at 60°C or above.
  - Make sure food handlers maintain appropriate levels of personal hygiene (e.g. washing hands thoroughly after going to the toilet).

Vibrio species

- *V. cholerae* is a major concern in countries where poor sanitation and hygiene lead to contamination of drinking water by human faeces. The gastroenteritis caused by this species is called cholera. Cholera may be fatal because the severe diarrhoea caused can result in significant loss of body fluids.
- Gastroenteritis caused by *V. parahaemolyticus* or *V. cholerae* is commonly associated with contaminated seafood that is eaten raw, inadequately cooked or has been adequately cooked then cross contaminated by raw seafood. The level of cells naturally present in seafood is

typically too low to cause illness, so adequate refrigeration after harvesting is critical to stop cells growing.

- Cases of *V. vulnificus* infection are rare and they usually involve individuals in vulnerable populations who have an underlying illness.
- Specific control measures:
  - Purchase seafood from reputable suppliers who harvest from safe areas.
  - Rapidly chill raw seafood to 5°C or below; this is particularly essential in summer.
  - Cook seafood using adequate time–temperature combinations (e.g. 70°C for 2 minutes), unless intended to be eaten raw.
  - Prevent cross contamination between ready-to-eat foods and utensils and other items used with raw seafood (e.g. storage and transport crates).
  - Do not serve raw seafood or shellfish to individuals in vulnerable populations.

Yersinia enterocolitica

- Illnesses caused:
  - Gastroenteritis is the primary illness, but *Y. enterocolitica* may also cause reactive arthritis.
- Growth conditions:
  - Can grow at refrigeration temperatures, although at a slow rate.
  - Optimum growth temperature is about 30°C.
- Specific control measures:
  - Cook raw meat and poultry using adequate time–temperature combinations (e.g. 70°C for 2 minutes) and avoid re-contamination after heating.
  - Store ready-to-eat products at 5°C or below for a limited time.

Overview of illnesses caused by foodborne viruses

The table below provides a summary of the viruses that are of primary concern in Australia and the illnesses they cause. Viruses do not produce toxins so only infections, not intoxications, occur.

Illnesses caused by foodborne viruses

Virus name or type	Typical incubation period	Illness and main symptoms
Hepatitis A virus	25 to 30 days	Hepatitis A: fever, tiredness, nausea, abdominal pain, jaundice, dark urine
Noroviruses	36 hours	Gastroenteritis: nausea, vomiting, diarrhoea, abdominal pain, headache, fever, tiredness
Astroviruses	1 to 2 days	Gastroenteritis: diarrhoea
Rotaviruses	1 to 3 days	Gastroenteritis: diarrhoea, vomiting, fever

Individual foodborne viruses – up close and personal

Noroviruses

- Illness caused:
  - Main cause of foodborne viral gastroenteritis worldwide.
  - Gastroenteritis lasts for 1–2 days; it is generally mild but it can be life-threatening to the elderly or those with weakened immune systems.
- Primarily spread through the faeces of infected individuals: either directly or via food or water contamination. Vomiting can also spread norovirus particles through the air. Noroviruses are highly contagious.
- Noroviruses are highly resistant to many chemical sanitisers, but chlorine (bleach) is effective if applied correctly after cleaning (see Chapter 3, Box 29).
- Specific control measures:
  - If an infected person vomits at a food business premises: discard any food that is not in well-sealed containers (unless it is to be heated to at least 85°C throughout); and thoroughly clean and then sanitise all exposed surfaces with chlorine, including the outer surfaces of food containers.
  - Food handlers diagnosed as infected with norovirus must be prevented from handling food until a doctor advises they are fit to do so.
  - Food handlers should be trained in and follow strict hygienic practices, especially washing hands thoroughly, particularly those who handle ready-to-eat foods.

- Fruits, vegetables and seafood should be purchased from reputable suppliers.

Rotaviruses and astroviruses

- Illness caused:
  - Rotavirus infection is the most common cause of gastroenteritis in children in developing nations; the illness can be quite severe, with those under 2 years most vulnerable.
  - Astrovirus infection generally causes diarrhoea, lasting 2–4 days.
- The spread to food and control measures are very similar to noroviruses, except rotaviruses and astroviruses are not commonly known to be spread by vomiting.

Hepatitis A virus

- Illness caused:
  - Common symptoms include fever, fatigue, jaundice, loss of appetite and vomiting.
- Primarily person-to-person spread through faecal matter from infected person. Hepatitis A is shed in faeces at high levels during the incubation period, and this starts even before those infected experience any symptoms. This is the 'danger period' because food handlers do not know they should stay away from work or normal duties.
- Can remain active in high-acid foods, and for long periods frozen or refrigerated
- Specific control measures are similar to norovirus. However, it is not commonly known to be spread by contamination caused by vomiting. Food handlers diagnosed with hepatitis A must stay at home until a doctor determines they are fit to return, because the virus can be shed for longer periods than other foodborne viruses.
- Hepatitis A vaccines are available and vaccination of food handlers is recommended. Any staff member who has come into contact with another person who has had hepatitis A should get vaccinated immediately.

Facts about foodborne illnesses and individual pathogenic microorganisms provided in this chapter were partially sourced from: Foodborne Microorganisms of Public Health Significance, Food Microbiology – an Introduction, and Microbiological Specifications of Food Pathogens (for details see pages 271–273).

Sources of information

The Food Standards Australia New Zealand (FSANZ) Food Standards Code

The Code is freely accessible via the FSANZ website:

- www.foodstandards.gov.au

go to the 'Food Standards Code' link under 'FOOD STANDARDS'.

To keep up to date with changes to the Code, you can join a free email subscription to the Food Standards Gazette and receive notification when amendments are made. To register, go to the 'Subscription service' link under 'QUICK LINKS' on the FSANZ website. Through this subscription service you can also receive notifications for Australian consumer-level recalls and other updates.

Standards from the Food Standards Code referred to in this book are listed in the table below.

Standard	Name
1.2.1	Application of Labelling and Other Information Requirements
1.2.2	Food Identification Requirements
1.2.3	Mandatory Warning and Advisory Statements and Declarations
1.2.4	Labelling of Ingredients
1.2.5	Date Marking of Packaged Foods
1.2.6	Directions for Use and Storage
1.2.9	Legibility Requirements
1.3.1	Food Additives
1.3.2	Vitamins and Minerals
1.3.3	Processing Aids

Standard	Name
1.4.4	Prohibited and Restricted Plants and Fungi
1.6.1	Microbiological Limits for Food
1.6.2	Processing Requirements
2.2.1	Meat and Meat Products
2.2.3	Fish and Fish Products
2.3.1	Fruit and Vegetables
2.6.3	Kava
2.9.1	Infant Formula Products
2.9.2	Foods for Infants
2.9.3	Formulated Meal Replacements and Formulated Supplementary Foods
3.1.1	Interpretation and Application
3.2.2	Food Safety Practices and General Requirements
3.2.3	Food Premises and Equipment
3.3.1	Food Safety Programs for Food Service to Vulnerable Persons
4.2.1	Primary Production and Processing Standard for Seafood
4.2.2	Primary Production and Processing Standard for Poultry Meat
4.2.3	Primary Production and Processing Standard for Meat
4.2.4	Primary Production and Processing Standard for Dairy Products

User guides to the Food Standards Code

The FSANZ website also contains several user guides, developed to help food manufacturers to interpret and apply the Code. To find the user guides, go to the 'User guides' link under 'FOOD STANDARDS'.

User guides referred to in this book are:

- Safe Food Australia (2001)
- User Guide to Standard 1.2.5 – Date Marking of Packaged Foods (2001)
- User Guide to Standard 1.6.1 – Microbiological Limits for Food with additional guideline criteria (2001)

Other publications

Name	Year	Author and/or source	Free?
Australian food statistics 2008	2009	Australian Government Department of Agriculture, Fisheries and Forestry (DAFF) Available on the DAFF website: www.daff.gov.au/agriculture-food/food/publications	Yes
Australian manual for control of *Listeria* in the dairy industry	1999	Australian Dairy Authorities' Standards Committee Available on the Australian Quarantine and Inspection Service website: www.daff.gov.au/aqis/export/dairy/pubs-guidelines	Yes
Cook chill for foodservice and manufacturing: guidelines for safe production, storage and distribution	2008	Brigitte Cox and Marcel Bauler Published by Australian Institute of Food Science and Technology (AIFST) Can be ordered on the AIFST website: www.aifst.asn.au	
Design, construction and fit-out of food premises (Australian Standard): AS 4674-2004	2004	Standards Australia Available via SAI Global website: www.saiglobal.com/Information/Standards	
Foodborne microorganisms of public health significance	2003	Australian Institute of Food Science and Technology (AIFST), NSW Branch, Food Microbiology Group Can be ordered on the AIFST website: www.aifst.asn.au	
Food industry guide to allergen management and labelling	2007	Australian Food and Grocery Council (AFGC) Available on the AFGC website: www.afgc.org.au/tools-guides-/-tools-a-guides.html	Yes
Food industry recall protocol: a guide to writing a food recall plan and conducting a food recall	2009	Food Standards Australia New Zealand (FSANZ) Available on the FSANZ website: www.foodstandards.gov.au/scienceandeducation/publications/	Yes

Name	Year	Author and/or source	Free?
Food microbiology: an introduction	2006	Tim Hutton Published by: Campden & Chorleywood Food Research Association, United Kingdom	
Food safety guidelines for the preparation and display of sushi	2007	NSW Food Authority Available on the NSW Food Authority website: www.foodauthority.nsw.gov.au/_Documents/industry_pdf/Sushi-Guidelines-Eng.pdf	Yes
Food Standards Agency guidance on the safety and shelf-life of vacuum and modified-atmosphere packed chilled foods with respect to non-proteolytic *Clostridium botulinum*	2008	UK Food Standards Agency Available on the Food Standards Agency website: www.food.gov.uk/multimedia/pdfs/publication/vacpacguide.pdf	Yes
Guidelines for the microbiological examination of ready-to-eat foods	2001	Food Standards Australia New Zealand (FSANZ) Available on the FSANZ website: www.foodstandards.gov.au/newsroom/publications/guidelinesformicrobi1306.cfm	Yes
Guidelines for the safe manufacture of smallgoods	2003	Meat & Livestock Australia Email: Publications@mla.com.au or phone (02) 9463 9124	
Guidelines on the verification of re-heating instructions for microwaveable foods	1997	UK Microwave Working Group Published by: Campden & Chorleywood Food Research Association, United Kingdom	
Food Safety: Guidance on skills and knowledge for food businesses	2002	Food Standards Australia New Zealand (FSANZ) Available on the FSANZ website: www.foodstandards.gov.au/scienceandeducation/publications/	Yes
Food Safety: Temperature control of potentially hazardous foods	2002	Food Standards Australia New Zealand (FSANZ) Available on the FSANZ website: www.foodstandards.gov.au/scienceandeducation/publications/	Yes

Name	Year	Author and/or source	Free?
The annual cost of foodborne illness in Australia	2006	Australian Government Department of Health and Ageing Available on the OzFoodNet website: www.ozfoodnet.org.au/internet/ozfoodnet/publishing.nsf/Content/reports-1	Yes
Listeria management program	2008	NSW Food Authority Available on the NSW Food Authority website: www.foodauthority.nsw.gov.au/_Documents/industry_pdf/listeria-management-program.pdf	Yes
Listeria and food – advice for people at risk		Food Standards Australia New Zealand (FSANZ) Available on the FSANZ website: www.foodstandards.gov.au/_srcfiles/Listeria.pdf	Yes
Microbiological specifications of food pathogens (Microorganisms in foods 5)	1996	International Commission on Microbiological Specifications for Foods Published by: Blackie Academic & Professional, United Kingdom	
Monitoring the incidence and causes of diseases potentially transmitted by food in Australia: Annual report of the OzFoodNet Network, 2007	2008	OzFoodNet Available via the OzFoodNet website: www.ozfoodnet.org.au/internet/ozfoodnet/publishing.nsf/Content/reports-1	Yes
Plastics materials for food contact use (Australian Standard): AS 2070-1999	1999	Standards Australia Available via SAI Global website: www.saiglobal.com/Information/Standards	
Vulnerable persons food safety scheme manual	2008	NSW Food Authority Available on the NSW Food Authority website: www.foodauthority.nsw.gov.au/industry/industry-sector-requirements/food-service-to-vulnerable-populations	Yes

Note: all website addresses were correct as at 10 March 2010

Useful contacts

Commonwealth, state and territory authorities

Australian Capital Territory Health Protection Service

Web: www.health.act.gov.au/c/health?a=sp&pid=1053918396

Phone: 02 6205 1700

Mail: Locked Bag 5, Weston Creek, ACT 2611

Email: hps@act.gov.au

Food Standards Australia New Zealand (FSANZ)

Web: www.foodstandards.gov.au

Phone: 02 6271 2222

Mail: PO Box 7186, Canberra BC, ACT 2610

New South Wales Food Authority

Web: www.foodauthority.nsw.gov.au

Phone: 1300 552 406

Mail: PO Box 6682, Silverwater, NSW 1811

Email: contact@foodauthority.nsw.gov.au

Northern Territory Department of Health and Families

Web: www.health.nt.gov.au/Environmental%5FHealth/Food%5FSafety

Phone: 1800 095 646

Mail: PO Box 40596, Casuarina, NT 0811

Email: envirohealth@nt.gov.au

Queensland Government Department of Health

Web: www.health.qld.gov.au/foodsafety/

Phone: 07 3234 0111

Mail: GPO Box 48, Brisbane, Queensland 4001

South Australian Government Department of Health

Web: www.dh.sa.gov.au/pehs/food-index.htm

Phone: 08 8226 7100

Mail: PO Box 6, Rundle Mall, SA 5000

Email: food@health.sa.gov.au

Tasmanian Government Department of Health and Human Services

Web: www.dhhs.tas.gov.au/health__and__wellbeing/public_and_environmental_health

Phone: 1800 671 738

Mail: GPO Box 125, Hobart, Tas 7001

Victorian Government Department of Health

Web: www.health.vic.gov.au/foodsafety

Phone: 1300 364 352

Mail: GPO Box 4057, Melbourne, Vic 3001

Email: Foodsafety@dhs.vic.gov.au

Western Australian Government Department of Health

Web: www.public.health.wa.gov.au/2/830/3/food_informatio.pm

Phone: 08 9388 4999

Mail: PO Box 8172, Perth Business Centre, WA 6849

Email: foodunit@health.wa.gov.au

Other government contacts

Australian Quarantine and Inspection Service (AQIS)

Web: www.aqis.gov.au

Phone: 1800 020 504

Mail: GPO Box 858, Canberra, ACT 2601

Safe Food Production Queensland

Web: www.safefood.qld.gov.au

Phone: 1800 300 815

Mail: PO Box 440, Spring Hill, Qld 4004

Email: info@safefood.qld.gov.au

How to find/where to go

Allergen information and tools for controlling

- **Allergen Bureau**

Web: www.allergenbureau.net

Phone: 1800 263 829

Email: info@allergenbureau.net

- **Australian Food and Grocery Council**

Web: www.afgc.org.au

Phone: 02 6273 1466

Email: info@afgc.org.au

Equipment and ingredients suppliers

- Look under 'Food preparation equipment' and 'Laboratory equipment supplies' in the phone book.
- Grab a copy of the Food & Drink Directory (Yaffa Publishing: Phone 02 9281 2333).
- Go along to the next food industry trade show in your area (e.g. Fine Food Australia).

Food Safety Programs – templates and software

- Look on your local state or territory authority website for templates and advice.
- Ask your relevant industry association.

Food Safety Auditors

- Refer to the Register of Approved Auditors on your local state or territory authority's website to find an auditor with appropriate accreditation.

Product Information Form (PIF)

- Available from the **Allergen Bureau** (see under 'Allergen information and tools for controlling' above).

Registered Training Organisations (RTOs)

- Contact your local state or territory authority.
- Search on the training.com.au website (www.training.com.au).

Testing laboratories and technical experts

- **National Association of Testing Authorities (NATA)** accredited labs (recommended):

Web: www.nata.asn.au

Phone: 1800 621 666

- For packaging advice contact the **Packaging Council of Australia**:

Web: www.pca.org.au

Phone: 03 9690 1955

Email: info@pca.org.au

- To locate an **AQIS Approved Persons** contact AQIS (see under 'Other government contacts' above).
- Look under 'Food technology consultants' in the phone book.

Glossary

Word or term	Meaning (relevant to food safety)
Acceptance criteria	Pre-established limits: these can be either a single point or a range of values. These are monitored to ensure equipment and processes are operating correctly. For example, checking that a cold room is between 2 and 5°C or the pH of a final product is below 4.6.
Aerosols	Tiny droplets of liquid that can travel long distances through the air. If these droplets contain pathogenic microorganisms, they can cause cross contamination.
Aseptic	A process where there is such tight control that no microbial contamination is expected.
Audit	A structured inspection of premises and/or documents to ensure that an appropriate system for controlling food safety hazards is in place and that the system is being followed.
Best-before date	The date that shows when food – if stored unopened in its original packaging and according to storage conditions on the label – will retain an acceptable level of quality.
Biofilm	A protective layer on surfaces that can protect microorganisms living within the layer from harsh conditions such as cleaning chemicals – this is similar to plaque that forms on teeth.
Botulinum cook	Heat processing food using sufficient heat and time to reduce the number of *Clostridium botulinum* spores that may be present to safe levels. When used in combination with hermetically sealed packaging, this allows low-acid foods to be safely stored at room temperature.
Calibration	Measuring the accuracy of readings given by a particular piece of equipment by comparing with readings given by another piece of equipment known to be accurate (e.g. reference thermometer). Thermometers, temperature gauges, pH meters and scales or balances all need regular calibration.
Cook chill product	Cook chill products are pasteurised then stored under refrigeration. The food is usually fully cooked and may only require re-heating by the consumer or may be eaten without any further heating.
Commercial sterility	Commercial sterility of heat processed food is achieved when sufficient heat is applied – sometimes in combination with control of pH and/or water activity – to kill all microorganisms capable of growing in the food at room temperature, and control microorganisms (including spores) of public health significance. Commercial sterility is only achieved if appropriate packaging methods are used.

Contact time	The length of time that a piece of equipment, surface or a food needs to be exposed to a sanitiser (or a disinfectant).
Contaminant	Microorganisms (or their toxins), chemicals and physical hazards that make their way into a food and cause it to be unfit to be eaten. This contamination can occur via other food, people, equipment and/or packaging.
Control measures	Actions that can reduce the risk of exposure to a hazard or can fully eliminate a hazard. For example, storing food at 5°C or below for a limited time.
Corrective action	An action taken to remove the cause of a problem and minimise the chance that it will occur again in the future. For example, increasing the frequency of maintenance of a piece of food processing equipment.
Cross contamination	Spread of a contaminant from one food or piece of equipment to another; for example, the transfer of pathogenic bacteria from raw food to cooked food because of poor food handling practices.
Food allergen	Naturally occurring proteins that can cause an abnormal and exaggerated immune response in sensitive individuals when eaten. Allergic reactions vary from mild illness (e.g. skin rashes) through to severe, life-threatening anaphylactic shock involving swelling of the airways, possibly leading to death.
Food handlers	People who directly engage in the handling of food, or handle (including clean) surfaces that are likely to come into contact with food. Handling of food includes making, manufacturing, producing, collecting, extracting, storing, transporting, delivering, preparing, processing, preserving, packing, cooking, thawing, serving or displaying of food.
Food hygiene	The practices used by food handlers to prevent cross contamination and foodborne illness. It includes practices used to maintain an adequate level of personal cleanliness such as washing hands thoroughly before handling food.
Food premises	A place from which a food business operates and food is prepared and/or stored. This can include a retail outlet, commercial kitchen, domestic kitchen, farm, vineyard or orchard.
Food recall	The removal from distribution, sale or risk of being eaten of food that may pose a health and safety risk to consumers.
Food safety management system	A documented system that requires a business to examine its operation and put controls into place to manage any significant hazards that are likely to occur; for example, a Food Safety Program.
Hazard	A potential source of harm. The three types of food safety hazards that may cause consumers harm are microbial, chemical and physical hazards.
Headspace	The area between the top surface of food in a package and the packaging, which contains air that is trapped in the package when it is sealed.

Heat resistant	The ability of microorganisms and toxins to survive or remain active after being exposed to a heat process, which would kill or inactivate non-heat resistant microorganisms and toxins.
Hermetically sealed	Packaging seals that are air and watertight, preventing microorganisms from entering the pack and contaminating the product after heat processing.
High-acid foods	Foods with a pH below 4.6, such as lemons and oranges.
High-risk foods	See **potentially hazardous foods**
Hurdles	Control measures that act as barriers to the growth or survival of microorganisms. One control measure (e.g. heating) may not be sufficient on its own, but, when combined with another hurdle (e.g. adding acid), the effect will be enhanced.
Infection	Pathogenic microorganisms growing or reproducing in the body after being eaten, which may or may not cause illness.
Kill-step	A process used to reduce any pathogens that may be present in food to safe levels.
Low-acid food	Foods with a pH above 4.6, such as red meat and vegetables.
Modified-atmosphere Packaging (MAP)	Replacing the air surrounding a food with a single gas (e.g. carbon dioxide) or mixtures of different gases (e.g. carbon dioxide plus nitrogen).
Monitoring	Periodically checking that factors critical to the safety of products are under control or operating effectively, and keeping a record of these checks. For example, checking that a cold room is running at 5°C or below every day.
MSDS (Material Safety Data Sheet)	Information provided by suppliers or manufacturers summarising a chemical's toxicity, health hazards, physical properties, fire and reactivity data. It also includes storage, spill and handling precautions.
Pasteurisation	Heating a product sufficiently to kill enough target pathogenic microorganisms so that any remaining are unlikely to cause foodborne illness, when combined with additional hurdles such as storing at 5°C or below for a limited time. The minimum time–temperature combination required to kill cells of pathogenic bacteria is 70°C for 2 minutes (or equivalent).
Pathogen	A microorganism that can cause illness in humans after it, or a toxin it has produced, is eaten.
pH	A measure of how much acid is in a food.
Post-process contamination	Microbial contamination that occurs after the product has received a heat process or other kill-step.

Potable water	Water of a quality that meets the Australian Drinking Water Guidelines and is suitable for drinking.
Potentially hazardous foods	Any food that may contain pathogenic microorganisms – either naturally occurring or through contamination during handling – and is able to support growth of, or toxin production by, these pathogens.
Raw materials	The basic components or ingredients used to make a product.
Ready-to-eat foods	Foods that would normally not be cooked or heated by the consumer before being eaten.
Risk	The likelihood of being exposed to a hazard and the likelihood of being harmed if exposed. For example, unpasteurised eggs are more likely to contain pathogenic bacteria than pasteurised eggs and people with weakened immune systems are at a higher risk of becoming ill if they eat them.
Sanitise	The process required to reduce the level of pathogens on food contact surfaces or equipment to safe levels. Very hot water, steam or chemical sanitisers (also called disinfectants) can be used.
Scale-up	Modifying the manufacturing process to convert it from small-scale to larger scale.
Shelf-life	The length of time a food can be stored under specified storage conditions without becoming a food safety hazard or developing unacceptable quality attributes.
Slowest heating point	The area in a package or piece of food that will take the longest to heat up. The location of the slowest heating point depends on the shape of the food or package and whether the food is a liquid, solid or somewhere in-between.
Spores	A form that some bacteria can take when environmental conditions are not in their favour. Spores can remain alive in a dormant state for many years. Then, if conditions become favourable for growth, they can be activated to form vegetative cells and grow.
Target microorganisms	Pathogens or spoilage microorganisms that are most likely to be a problem in a particular food and must be killed or controlled.
Temperature control	Maintaining the temperature of a food so any pathogens that may be present are not able to grow or produce toxin to high enough levels to cause illness. This requires that potentially hazardous foods remain at 5°C or below, 60°C or above, or at another temperature for a limited time if this can be proven to be safe.
Temperature danger zone	This is the temperature range between 5°C and 60°C where pathogens are able to grow best.

The Code	This refers to the Food Standards Australia New Zealand (FSANZ) Food Standards Code.
Thermal shock	The shattering or breaking of glass containers when they are heated or cooled too quickly.
Time–temperature combination	Heating food to a specific temperature during cooking or heat processing and holding it at this temperature for a specific minimum time.
Toxin	A chemical substance produced by pathogenic microorganisms that can result in illness. Toxins can be produced in the food before it is eaten or in the body after the food containing the pathogens is eaten.
Under processing	When the required time–temperature combination is not met during heat processing, possibly resulting in unacceptable survival of pathogens.
Use-by date	The form of date marking that must be applied to all products that need to be eaten by a certain date for health and/or safety reasons. It is illegal to sell food once the use-by date applied by the manufacturer has passed.
Vacuum packaging	Removing the majority of air from around food that is stored in airtight packaging. The main purpose is to exclude oxygen, making it more difficult for some spoilage microorganisms to grow and/or to reduce chemical changes that may limit shelf-life.
Validation	Gathering evidence that the control you have put into place to prevent food becoming hazardous will work. For example, checking that a specific time–temperature combination used during heat processing will actually reduce contamination by a target pathogen to an acceptable level.
Vegetative cells	Bacteria forms that are capable of growing if conditions are favourable.
Verification	Making sure that a food safety management system is being implemented correctly, by conducting checks such as reviewing records for completeness and checking that appropriate corrective actions have been implemented when necessary.
Vulnerable person	A person with a higher chance of developing foodborne illness and who may suffer more severe effects or complications, such as the very young, the elderly and those whose immune systems are seriously weakened. Pregnant women are also a vulnerable population.
Water activity (a_w)	The term used when describing the water availability in food. The water activity scale extends from zero (bone dry) to one (pure water).

UNIVERSITIES AT MEDWAY LIBRARY